ROUTLEDGE LIBRARY EDITIONS:
NUCLEAR SECURITY

Volume 27

ETHICS AND
NUCLEAR DETERRENCE

ETHICS AND NUCLEAR DETERRENCE

Edited by
GEOFFREY GOODWIN

LONDON AND NEW YORK

First published in 1982 by Croom Helm

This edition first published in 2021
by Routledge
2 Park Square, Milton Park, Abingdon, Oxon OX14 4RN

and by Routledge
52 Vanderbilt Avenue, New York, NY 10017

Routledge is an imprint of the Taylor & Francis Group, an informa business

© 1982 Council on Christian Approaches to Defence and Disarmament

All rights reserved. No part of this book may be reprinted or reproduced or utilised in any form or by any electronic, mechanical, or other means, now known or hereafter invented, including photocopying and recording, or in any information storage or retrieval system, without permission in writing from the publishers.

Trademark notice: Product or corporate names may be trademarks or registered trademarks, and are used only for identification and explanation without intent to infringe.

British Library Cataloguing in Publication Data
A catalogue record for this book is available from the British Library

ISBN: 978-0-367-50682-7 (Set)
ISBN: 978-1-00-309763-1 (Set) (ebk)
ISBN: 978-0-367-53689-3 (Volume 27) (hbk)
ISBN: 978-1-00-308291-0 (Volume 27) (ebk)

Publisher's Note
The publisher has gone to great lengths to ensure the quality of this reprint but points out that some imperfections in the original copies may be apparent.

Disclaimer
The publisher has made every effort to trace copyright holders and would welcome correspondence from those they have been unable to trace.

Ethics and Nuclear Deterrence

EDITED BY GEOFFREY GOODWIN

CROOM HELM
London & Canberra

© 1982 Council on Christian Approaches to Defence and Disarmament
Croom Helm Ltd, Provident House, Burrell Row,
Beckenham, Kent BR3 1AT
Reprinted 1983

British Library Cataloguing in Publication Data

Goodwin, Geoffrey
    Ethics and nuclear deterrence.
    1. Deterrence (Strategy) — Moral and religious aspects
    2. Atomic warfare — Moral and religious aspects
    I. Title
    172'.42   BR115.A85

ISBN 0-7099-1159-9

The Council on Christian Approaches to Defence and Disarmament (CCADD) comprises a multi-denominational group of Christians with differing responsibilities; governmental and non-governmental. The Council seeks to bring an ethical viewpoint to bear on disarmament and defence and related issues of national and international security. Its members meet for regular discussions and conferences at which a wide range of views is expressed.

The opinions expressed in this publication are the responsibility of the authors alone.

Printed and bound in Great Britain by
Billing and Sons Limited, Worcester

CONTENTS

List of Abbreviations

Foreword *The Archbishop of Canterbury*

Introduction — 13

1. Deterrence and Détente: the Political Environment
   Geoffrey Goodwin — 16
2. Theological Method in the Deterrence Debate *G.R. Dunstan* — 40
3. A Christian Unilateralism from a Christian Background
   Bruce Kent — 53
4. In Defence of Deterrence *Arthur Hockaday* — 68
5. Deep Cuts are Morally Imperative *Barrie Paskins* — 94
6. The Case for Negotiated Disarmament *Roy Dean* — 117
7. International Humanitarian Law: Principles and Practices
   Geoffrey Best — 143
8. Concluding Comments *Ronald Hope-Jones* — 166

Appendix: Excerpts from Some of the Church Statements Referred to by Contributors — 185

Index — 194

ABBREVIATIONS

ABM	anti-ballistic missile
ASAT	anti-satellite warfare
ASEAN	Association of South East Asian Nations
CBM	confidence-building measures
CCADD	Council on Christian Approaches to Defence and Disarmament
CD	Committee on Disarmament (UN)
CND	Campaign for Nuclear Disarmament
CSCE	Conference on Security and Co-operation in Europe
CTB	Comprehensive Test Ban
CW	chemical weapons
END	European Nuclear Disarmament
GCD	General and Complete Disarmament
IAEA	International Atomic Energy Agency
ICBM	inter-continental ballistic missile
ICRC	International Committee of the Red Cross
IISS	International Institute for Strategic Studies
ISMA	International Satellite Monitoring Agency
MBFR	mutual and balanced force reductions
MIRV	multiple independently-targeted re-entry vehicle
NATO	North Atlantic Treaty Organization
NNWS	non-nuclear-weapon state
NPT	Non-Proliferation Treaty
NWFZ	Nuclear-weapon-free zone
NWS	Nuclear-weapon state
OAS	Organization of American States
OAU	Organization of African Unity
PTBT	Partial Test Ban Treaty
RDF	rapid deployment force
RW	radiological weapons
SALT	Strategic Arms Limitation Talks
SCM	Student Christian Movement
SIPRI	Stockholm International Peace Research Institute
SLBM	submarine-launched ballistic missile
TNF	theatre nuclear forces
UNEF II	United Nations Emergency Force II
UNIFIL	United Nations Interim Force in Lebanon
UNSSD	United Nations Special Session on Disarmament

FOREWORD

The preservation of peace — a word that is at the heart of the Gospel — is the greatest of the challenges facing mankind today. It is a task to which we must bend all our energies, for the price of failure is too horrible to contemplate.

Men and women all over the world are deeply puzzled about how to achieve the objective which we all seek. There are continuing differences of view on the best way to proceed. These differences are reflected within the Church as well as within the community at large.

In these circumstances we have a duty to listen to each other, to use our minds as well as our hearts, and to wrestle with the deep ethical problems which these issues involve. Peace will not be won by superficial or sloppy thinking.

I welcome this book and hope that it will be widely read. Its authors reflect some of the different approaches which exist, but share the same aim. It is a serious work on a desperately important subject. If it helps people to think deeply and constructively about these matters — and I believe it will — the book will perform a most valuable service at a time when this task is becoming increasingly urgent.

INTRODUCTION

This collection of essays stems from a request to the Council on Christian Approaches to Defence and Disarmament (CCADD) by the British Council of Churches to produce a revised version of *The Search for Security: A Christian Appraisal*, published in 1973 by the SCM Press, but now out of print. That report, which was the product of a group composed mainly of CCADD members, was wide-ranging and examined some of the root causes of international conflict as well as the machinery for dealing with international disputes and what are customarily regarded as defence and disarmament issues. The CCADD group, convened under the chairmanship of Ronald Hope-Jones to revise the report, concluded early on that something more than the updating of the previous material was required. As the group's discussions continued, it also became clear that not only had the deterioration in East-West relations since the original report made consideration of the nuclear dimension of these relations more urgent, but most of the group were preoccupied with the ethical issues raised, particularly for Christians, by deterrence policies and by the debate between proponents of the multilateralist/unilateralist approaches to disarmament and arms control.

Consequently the group agreed to concentrate on these issues. Given the wide range of views represented in the group it seemed improbable, however, that sufficient of a consensus could be reached within the time available to make a group report worthwhile. It was therefore decided instead to commission a series of linked essays on the theme of 'Ethics and Nuclear Deterrence' which would examine the theological and ethical issues involved in nuclear deterrence and the unilateralist and multilateralist approaches to arms control and disarmament, but which would also include an essay on attempts to mitigate the appalling brutality of the many 'conventional' wars since 1945. An appendix gives extracts from some of the major Church pronouncements referred to in the essays. Whilst each of the essays remains the responsibility of the individual author, each has benefited from the early group discussions held under Ronald Hope-Jones's chairmanship, the detailed discussions of drafts at meetings of the group in 1981 under the chairmanship of Geoffrey Goodwin, and from the written comments of individual members of the group. As a result every contri-

14 *Introduction*

butor has tried to take account of the often widely differing views represented on the group, so that the essays will reflect, it is hoped, the continuing dialogue between its members. The group have appreciated the help of Mark Venables, Director of CCADD, who has also assembled the texts in the appendix.

The members of the group actively engaged in consideration of the essays have been as follows. (The names below preceded by an asterisk are those of contributors to this book.)

Sydney D. Bailey	Chairman of the Council of CCADD.
James Barnes	Deputy Chief Scientific Adviser (Projects), Ministry of Defence.
*Geoffrey Best	Professor of History and Dean of the School of European Studies, University of Sussex.
*Roy Dean	Director, Arms Control and Disarmament Research Unit, Foreign and Commonwealth Office.
Michael Donelan	Senior Lecturer in International Relations, London School of Economics and Political Science, University of London.
*Canon G.R. Dunstan	F.D. Maurice Professor of Moral and Social Theology, King's College, University of London.
*Geoffrey Goodwin	Emeritus Professor of International Relations, London School of Economics and Political Science, University of London.
*Sir Arthur Hockaday, KCB, CMG	Second Permanent Under-Secretary of State, Ministry of Defence.
*Ronald Hope-Jones, CMG	Retired diplomat; Head of the Arms Control and Disarmament Department, Foreign and Commonwealth Office, 1967-70.
Frank Judd	Director, Voluntary Service Overseas; Parliamentary Under-Secretary of State for Defence (Navy) 1974-6. Minister of State for Foreign and Commonwealth Affairs, 1977-9.
Elizabeth Young (Lady Kennet)	Journalist and author of *Farewell to Arms Control?* (Harmondsworth, 1972).
*Bruce Kent	General Secretary, Campaign for Nuclear Disarmament.
*Barrie Paskins	Lecturer in War Studies and formerly Defence Lecturer in the Ethical Aspects of

Introduction

John Roper MP	War, King's College, University of London. Opposition front-bench spokesman on defence, 1979-81. Chief Whip, Social Democratic Party, 1981.
Mark Venables	Director, CCADD.
The Revd Hamish Walker	General Secretary, Fellowship of Reconciliation.

Members of the group holding official appointments have taken part in their personal capacity; similarly the essays by Sir Arthur Hockaday and Roy Dean express purely personal viewpoints and should not be taken as necessarily reflecting those of Her Majesty's Government.

Most books touching upon issues of nuclear deterrence usually express one particular point of view, either for or against a deterrent posture, and are often polemical in tone. We believe that this collection brings together an informed, carefully considered and representative cross-section of Christian viewpoints on the dilemma posed by the existence of nuclear weapons, ranging from the view that a deterrent posture is morally defensible for Christians on prudential grounds as the most reliable means, however detestable, of averting nuclear war, to the conviction that nuclear weapons are 'an offence to God and a denial of his purpose for man' and that as a matter of principle every effort should be made to diminish and eventually eliminate reliance on these weapons, if need be by unilateral nuclear disarmament. Ronald Hope-Jones brings out well in his 'Concluding Comments' the opposing arguments as well as the area where there is a measure of consensus. As will be evident, these essays are not intended to provide answers to the many perplexing problems they raise. Their aim is to contribute to a more informed and searching debate, particularly but not only amongst Christians, about these problems. A further hope is that they may contribute, however indirectly, to the shaping of policies most likely to avoid the horrors of nuclear devastation, but without sacrificing Western civilisation.

1 DETERRENCE AND DETENTE: THE POLITICAL ENVIRONMENT

Geoffrey Goodwin

Deterrence and *détente* have been two of the dominant concepts of East-West relations for well over a decade; indeed, the analogous concepts of a *pax atomica* and of a 'thaw' in US-Soviet relations go back to the mid-1950s. In the West both concepts are now under challenge. The threat of a new momentum to the nuclear arms race has given rise to a renewed tide of nuclear pacificism, while from the other pole of the political spectrum détente is denounced as a dangerous distraction from the drive towards nuclear modernisation. Official pronouncements may claim them to be complementary tasks of Western policy, but in practice deterrence and détente are more often than not seen as inherently contradictory.

The argument in this essay is that the preservation of a relationship of 'mutual deterrence' between West and East (as defined by Arthur Hockaday in Chapter 4,[1] that is when 'each side possesses sufficient missiles to survive a first strike from the other and to mount a retaliatory strike which would cause unacceptable damage to the aggressor') must be matched by a determined effort to reduce the level of tension between them and to resolve their more dangerous mutual differences by negotiation, that is by a process of détente. It also assumes that, although the ingredients of the relationship have changed over time, the maintenance of a relatively robust central balance of deterrence (in the sense in which Barrie Paskins writes in Chapter 5)[2] has so far effectively dissuaded either superpower from pressing its challenge to the point where nuclear war might have become unavoidable, in the realisation that in such a war no possible gains could outweigh the appalling losses both sides would suffer. Despite American nuclear predominance at the time, the Cuban missile crisis of 1962 provides the most striking instance of the risks of a nuclear war persuading the superpowers to draw back from the brink; over West Berlin and the recurrent Middle East conflicts and possibly in successive Eastern European and South East Asian crises those risks would also seem to have persuaded them to act with greater constraint. Nevertheless, the continued stability of the central nuclear balance is not to be counted upon indefinitely, while many strategists contend that the nuclear relationship at the European theatre level has recently been moving perceptibly against the West. With the increased friction in East-West relations during the last few years the prospects for détente and for meaningful measures of arms control also look distinctly bleak. Yet it is precisely in this kind of situation that both become more necessary than ever.

This essay will in the first part attempt to identify those sources of tension in the world at large which threaten to exacerbate the diplomatic climate generally and East-West relations in particular. It will then turn to a brief analysis of Soviet capabilities, intentions and ideology and the different kinds of threat they may be thought to pose to Western interests and values. To look at the other side of the picture, this part will conclude with a quick look at the threats from the West that may be perceived by Soviet leaders. The second part of the essay will examine the extent to which it is still realistic to work for a meaningful measure of détente between the major power blocs, especially in the area of arms control, the belief being that considerable risks can be taken to reduce the present level of nuclear arsenals without necessarily upsetting the relationship of mutual deterrence. The third part will touch upon some wider-ranging issues on the role of principle and prudence in the management of international politics in a nuclear age.

The essay is written in what, for want of a better term, I call a spirit of Christian realism. I spell this out further in the concluding part. Briefly, it affirms the relevance to political behaviour of a transcendental Christian ethic and of the principles to be derived from it, but it recognises that the application of these principles to the political realm is bound to be partial and relative, in the sense that 'universal moral principles cannot be applied to the actions of states in their abstract universal formulation . . . they must be, as it were, filtered through the concrete circumstance of time and place'.[3] To posit the ideal, to highlight the relevant principles is a worthy Christian calling. To help identify, as is attempted in this essay, the political realities which are crucial to the attainment of even an approximation to the ideal and also the political processes through which the realisation of the principles must be sought is, I would claim, no less a Christian responsibility.

I

Order and Tension in International Society

The whole notion of an international society, precarious as it has always been, is under attack. That society, made up as it is of sovereign states with often widely differing interests and lacking any overall system of government, is necessarily in large measure an anarchical society, in which states are inevitably moved by a sense of insecurity. Yet the prospects for a more acceptable level of security internationally, and hence of avoiding nuclear war, depend in large measure on the successful resistance to that attack and on the acknowledgement by a majority of sovereign states, especially by the most powerful, that they belong to an international society in which their conduct is to be governed not solely by the facts of power, or by the imperatives of their own ideology, but also by a generally accepted set of rules, conventions and norms of behaviour for the more orderly management of their relations.

The growing interdependence of states is often said to make for a readier acceptance of this notion of an international society — and of the obligations that stem from it. The disciplines of the central nuclear balance, the spreading ethos of economic and social modernisation, the communications revolution — the world as a 'global village' — have conditioned perceptions, it is claimed, to the point where the idea of an international society is widely taken for granted. This is a comforting and not altogether unwarranted view, but for the most part it is not the one taken here. The other side of the coin is that in the last forty years the world has seen a steady process of political fragmentation with the coming to independence of former West European imperial possessions, a growing cultural pluralism as indigenous cultures have broken through an often thin Western veneer, the widespread reversion to the seamier aspects of Renaissance diplomacy and to a frequent disregard for diplomatic proprieties, a marked disdain for many well-established tenets of international law as a relic of a colonial past or of a bourgeois-dominated international society, and an often appalling disregard for the rights of the weak. Modernisation on Western lines has in numerous cases heightened internal tensions between a crumbling traditional order and often socialist-minded modernising elites and has contributed to a resurgence, in areas of particular interest to the West, of a militant Islamic fundamentalism which threatens to exacerbate local tensions and to sour external relationships. One result of interdependence is the increasing vulnerability of states, both great and small, to external pressures, a vulnerability particularly acute for many Western countries heavily reliant for their livelihood on world markets and on essential fuel and mineral imports. Another is that the shock waves of what could previously be regarded as mainly local conflicts are now more likely to have world-wide reverberations and to present opportunities for competitive external intervention.

This is a depressing picture. Even if today there still is amongst the majority of states more than a tenuous regard for diplomatic proprieties and legal obligations, and some deference to the inhibitions on the use of force which spring not merely from fear of a nuclear holocaust, the achievement by the Soviet Union of effective world status and an American determination to reverse what is seen as a decline in American power and resolve, together with the increased turbulence in so much of the so-called Third World, has helped to exacerbate existing rivalries to the point where the whole idea of an international society is at risk.

It should not, of course, be imagined that the existing system of order in international society can be frozen for all time by arresting the processes of change or by denying a Soviet Union with expanding world interests all opportunities for extending its influence. One of the diplomatic arts is to discern the changing configurations of power amongst the more powerful and to work for their more orderly management, whether jointly or collectively, whilst at the same time responding to the interests of the less powerful and to the need to remedy the more glaring injustices in international society, the persistence of which may threaten the very international order it is desired to uphold. This task of managing change is not the prerogative of any single power, or 'world policeman'; it is one for all the major powers — including both the superpowers — acting wherever practical jointly and where possible collectively through the United Nations and particularly the Security Council, the potentialities of which for the relatively peaceful management of change are only too easily overshadowed by the often meaningless rhetoric of the General Assembly. Nevertheless, it has to be admitted not only that East-West rivalries often inhibit the scope for joint or collective action, but that the leverage the major powers can exercise over recalcitrant lesser powers, even when acting through the Security Council, has considerably diminished. Consequently the task of managing more localised tensions and disputes may fall increasingly on regional bodies such as the Organization of American States (OAS), the Organization of African Unity (OAU), the Arab League and the Association of South East Asian Nations (ASEAN). Despite considerable internal dissension and the external pressures to which they are often subjected, it is to the strengthening of such bodies that the West should look.

Change is not only inherently disruptive, but it can provide opportunities for external interventions when superpower rivalries are superimposed upon local conflicts. External intervention does not necessarily involve force. Much of the time it is limited to diplomatic and other forms, including economic instruments and, above all, to the provision of arms and covert subversion. Where it has involved force the costs for the major powers concerned have often been high, whether at Suez (1956), in Vietnam or in Afghanistan. On the other hand, the risks of retaliatory military action by a competing major power in such cases have been reduced by the probability that it would escalate a crisis to the nuclear level. This applies even more where the superpowers consider vital security interests are at stake. Even though means of dissuasion short of the threat, let alone the overt use of force, can still be quite effective (for example, the West's warnings against Soviet intervention in Poland have been a not insignificant inhibition), events in 1956 and 1968 in Eastern Europe highlighted the risks in a nuclear age of any form of armed retaliation by the West to check Soviet reassertion of the 'solidarity' of the socialist camp against the evident wishes of the Hungarian and Czechoslovak peoples for greater freedom. Mutual acceptance of what is euphemistically called the superpowers' respective 'spheres of responsibility' may seem to condone serious injustices of this kind but it at least reduces the risks of nuclear confrontation and, as Gordon Dunstan points out in Chapter 2, there are sometimes sound moral reasons for putting security before justice.

A related difficulty is that there is a good deal of confusion about the kinds of intervention that can be condoned and those that should be condemned. Thus, the Soviet invasion of Afghanistan has been almost universally condemned. By contrast India's interventions in East Pakistan or Tanzania's intervention in Uganda were generally condoned; the overthrow of the barbaric Pol Pot regime in Kampuchea by Vietnam forces evoked considerable sympathy despite diplomatic criticism; the use of force by the Patriotic Front in Zimbabwe was widely supported, as has been the use of force by liberation movements in Southern Africa generally. Even the legal position is unclear. Many argue that Article 2 (4) of the UN Charter[4] gives *carte blanche* to those claiming to be acting against a racist regime in furtherance of the concepts of justice enshrined in the Purposes of the Charter; others claim that this is to put a misleading construction on an article intended to outlaw the threat or use of force except in self-defence. Despite these contradictory arguments one generally accepted instrument for checking unilateral interventions in conflict areas of particular sensitivity to the superpowers has been UN peace-keeping forces. Admittedly experience is mixed; but UN forces have in several cases provided more effective

Deterrence and Détente 21

'buffers' between contestants than have those of regional bodies. The United Nations Emergency Force II (UNEF II) created an interesting precedent by being able to call upon sophisticated American surveillance and monitoring devices to check infringements of the demilitarised zone. Some kind of UN force has been seen as an integral ingredient of any Namibian settlement and it is not impossible that a strengthened United Nations Interim Force in Lebanon (UNIFIL) might place some constraint upon the prolonged civil war in Lebanon and upon the recurrent clashes between Palestine Liberation Organization (PLO) and Israeli forces. UN forces tend to 'freeze' situations and do not necessarily make a more lasting settlement easier. Nevertheless, to neglect the potential of the UN as an instrument of peace-keeping simply because the organisation is no longer quite so amenable to Western influence might merely play into the hands of those intent on exploiting opportunities for external intervention for their own ends in a way which would threaten a further deterioration in the diplomatic environment and make the pursuit of détente even more difficult.

Power and Ideology in East-West Relations

The build-up of Soviet military strength; Soviet achievement of one of the hallmarks of a world power, namely the capacity to project its power, either directly or by proxy, into virtually every part of the globe; and the apparent confirmation of the inherently expansionist nature of Soviet policy by the Soviet incursion into Afghanistan, have all contributed to a sharp deterioration in East-West relations and have resurrected fears of a Soviet threat akin to those held at the height of the cold war. In Europe NATO forces have to face Soviet conventional forces which are not only superior numerically, but are now probably a match qualitatively for their own. Moreover the rapid Soviet deployment of medium-range nuclear delivery systems (for example the Backfire bomber and the SS.20 missile) and the considerable Soviet capability for offensive chemical warfare are seen as possibly tipping the theatre balance decisively in the Soviet Union's favour. In Western eyes this Soviet build-up in Europe would seem not only to exceed any rational requirement for defence, but also to bear all the marks of an offensive military capability. Hitherto the Soviet edge in conventional forces in Europe has been offset by United States superiority at the level of the strategic nuclear balance. However, over the last few years the position has moved from one of unquestioned American superiority to 'essential equivalence', while the declining survivability of US intercontinental ballistic missiles (ICBM) silos has led to the notion of the 'window of opportunity', according to which the Soviet Union could be tempted to launch diplomatic, political and even military offensives around the world in the early or middle 1980s. This has commonly been regarded as the most important national security problem facing the US at the outset of the 1980s.[5] This needs to be taken with a pinch of salt. As the International Institute for Strategic Studies (IISS) *Strategic Survey, 1980-1981* points out, the vulnerability of the US ICBM is much more a theoretical than an operational concept, and Soviet superiority in ICBM is balanced by the US superiority in both the submarine-launched ballistic missile (SLBM) and bombers.

Therefore, if the Soviet Union were able to make political gains during the period of the 'window of opportunity', this would not be because of any objective strategic balance in her favour, but rather because the limited perspectives of the American strategic debate has led to an underestimation of the conditions of nuclear deterrence and of US capabilities.[6]

And, I would add, to a typical overestimation of Soviet nuclear capabilities.

Nevertheless the shift in the European theatre balance would appear to be more serious, and has inevitably accentuated fears that the Soviet Union might use its superiority at both the conventional and nuclear levels not so much for an armed incursion into Western Europe but rather to exercise pressure either upon West Berlin or upon the European members of NATO in such a way as to loosen the ties that bind the United States and its European allies together. If this were to happen there might be considerable domestic pressure in the United States to reduce the level of American forces in Europe while the very credibility of the American nuclear commitment might be so called into question as to threaten the dissolution of the NATO alliance. And it is the dissolution of that alliance which is still seen by many in the West as one of the prime objectives of Soviet policy.

The implications of even a loosening of NATO ties are certainly disquieting. NATO is, it is true, only an alliance, a coalition of states, and there are bound to be strains between members, particularly as the relative economic and political influence of the West European members increases; and there may at times be considerable friction between individual members, as there has been between Greece and Turkey. Nevertheless NATO is still very much an expression of the old adage 'hang together or hang separately'. Moreover, it is the vehicle through which American ground troops and weapons are stationed in Western Europe and an important cement of the American nuclear commitment to the defence of Western Europe. And it provides additional channels through which the European members can hope to influence American thinking and policies. NATO also can help to ensure that negotiations with the Soviet Union on a wide range of issues — whether over West Berlin or mutual and balanced force reductions (MBFR) — are conducted from the basis of an agreed negotiating position which all have a share in formulating.

Formally speaking, however, the area of NATO's concern is limited to Europe and the Mediterranean and the North Atlantic Ocean north of the Tropic of Cancer. Yet apart from the still anomalous and precarious position of West Berlin, the main Soviet threats in Western eyes are perceived as lying outside the NATO area. The Soviet invasion of Afghanistan certainly provided the West with a means of establishing somewhat closer ties with much of the Islamic world, but it removed Afghanistan from its traditional role as a buffer state, bringing the Russians nearer to the Gulf states and the West's essential oil supplies. That Soviet access to the oil supplies and warm water ports of the Gulf was almost certainly not the chief motive behind the Soviet occupation of Afghanistan[7] is not, strictly speaking, directly relevant. The fact is that, in addition to the worrying implications of the Soviet military presence in Ethiopia and South Yemen, the strategic picture in this area has radically changed. Without the withdrawal of troops from Afghanistan and the strengthening, particularly internally, of the latter's immediate southern neighbours (particularly Pakistan with its secessionist-minded Baluchistan), perceptions of a serious threat to Western interests in the area are bound to persist. The emphasis on 'low-profile' political and economic diplomacy directed to reducing the risks of subversion and insurrection in the region is clearly right. A rapid deployment force (RDF) 'over the horizon' might conceivably act as a boost to the morale of friendly regimes in the Gulf states and a symbolic warning to the Soviet Union. Nevertheless, the logistical problems are formidable and if, as I believe, the

> main threat is not a Soviet invasion or an attack by a Soviet proxy but internal instability or a domestic change of policy . . . its presence in an unstable area could aggravate internal turbulence, turn the opposition into a shrilly anti-American or anti-Western direction and tempt the Western powers into using the available force to control events, which might be a fatal mistake.[8]

I would go further. If the fighting capability of the force is very limited, as is almost certain to be the case, its operational involvement in the area could, if maladroitly handled, put it so at risk that only a nuclear threat could retrieve it from a desperate situation. Even apart from the RDF, the dangers of pouring military hardware into the Gulf states with little thought to the stability of the recipient regimes have been highlighted by events in Iran and elsewhere. It must also be emphasised that for most of the Arab world a Soviet threat is a distant one and is quite secondary to what is seen as the West's, and especially the United States', equivocacy about the rights of the Palestinian peoples to self-determination and of the right of the PLO to be treated as their legitimate representative. Adroit diplomacy is what is called for, not a flexing of military muscles.

Much the same might be said of reactions to the so-called Soviet global threat consequent upon its achievement of a world-reach comparable to that of the United States. Perceptions of this 'threat' are strongly coloured by a sense of psychological shock amongst many Americans who had long regarded the United States as the only truly world power. In fact, Soviet experience has been distinctly mixed; the Soviet Union was expelled from both Egypt (1972) and Somalia (1976-7), Vietnam is proving a troublesome ally in South East Asia, the Cuban involvement in Angola is a costly drain on Soviet resources, as is the provision of Soviet arms and supplies to Ethiopia, while Soviet forces are still bogged down in Afghanistan. Nevertheless, for Soviet leaders one of the main attractions of the new-found capacity to project Soviet power throughout the world is the extent to which it enhances Soviet prestige and gives credence to their assertion that the Soviet Union now has the right, just as much as has the United States, to be consulted on all major world issues. Moreover, it must be expected that although the Soviet Union appears to have asserted its influence or its presence more as opportunities have arisen than according to some grand design, it is bound to be tempted as a matter of principle to support 'progressive' forces in countries of the so-called Third World, if necessary with military assistance as well as military hardware and diplomatic backing. Not to do so would be a disservice to the 'proletarian internationalism' of which the Soviet Union still sees itself as the legitimate leader. One reputable scholar has indeed suggested that

> the Soviet Union may, in the 1980s, be increasingly tempted to compensate abroad for domestic weaknesses and tensions: declining

growth, serious economic inefficiencies, one or two succession periods, a growing need for oil from the outside, changes in the demographic composition of the Soviet Union — all may lead to a quest for external triumph.[9]

In particular, widespread instability in the Third World, as well as the potential for growing economic friction with the West, could provide additional opportunities for the extension of Soviet influence into regions on which the West is heavily dependent for a wide range of essential non-fuel mineral supplies. Despite the shift in mineral exploration towards North America and Australia, Africa south of the Sahara supplies the West with many important minerals, including some vital to their arms industries. Import dependency does not, of course, necessarily denote vulnerability in the sense of a direct threat to the economic livelihood of importing countries, let alone to the national security. Yet the dividing line between dependence and vulnerability is often pretty thin — especially when supplies have to be brought a considerable distance by sea — so that assurance of necessary supplies of key materials, whether through stockpiling, switching to less vulnerable sources, or covert forms of control, is a necessary element in security calculations.

It is foolish to dogmatise on these issues. Nevertheless, whilst not denying the need to retain a minimal second-strike nuclear capability (that is one which could survive a first strike and mount a retaliatory strike which would cause unacceptable damage) to ward off threats of nuclear blackmail in potential flash-point situations (for example West Berlin, the Gulf, an Arab-Israeli war), the debate on nuclear deterrence in other respects seems to be pretty irrelevant to those parts of the world which lie outside the NATO area and in which Soviet interests are most likely to clash with those of the West. The task of shoring up defences against Soviet expansion here is one *mainly* for diplomacy, and particularly for economic diplomacy, to demonstrate that it is to the West that these countries can look not simply for defence equipment but principally for help in the achievement of their plans for social and economic development, progress on which is often essential to the viability of their countries and their present regimes. In this respect the West has a distinct edge for, although the Soviet Union may have achieved a world-reach and become a major supplier of military hardware, its political image is already badly tarnished and it has lacked the economic resources to make a significant contribution even to the economic needs of those African countries (Ethiopia, Angola and Mozambique) to which it has been most committed.[10]

Soviet Perceptions

Any appreciation of the political environment should seek to portray it as it looks from 'the other side'. The obverse of the picture of the West's perceptions of Soviet threats is the Soviet Union's perceptions of the challenges their country may have to face over the next decade or so. Although Soviet leaders may perceive the correlation of forces at the world level as having moved to their advantage, the picture is in many other respects, both internally and externally, very bleak.[11] For the Soviet Union the Sino-American *rapprochement* coupled with vehement Chinese denunciations of Soviet 'hegemonic' pretensions holds out the possibility in the not too distant future of a developing triangular central balance weighted significantly against the Soviet Union. Indeed, the emergence of China, with at least the potential of a superpower, from her self-imposed isolation and her assiduous cultivation of normal relations – both with the major West European countries and with Japan – in her search for rapid modernisation is bound to raise disquieting questions for the Soviet leaders. It is true that, given the demands of modernisation, China may be neither psychologically nor materially ready to play a leading role in the near future. But the potential is there and there is an inclination in Washington to 'play the China card' by explicitly exploiting the Sino-Soviet rift and the need for Soviet military contingency planning to be geared to a two-front strategy. Too brash an attempt to do so would almost certainly enhance the Soviet Union's sense of insecurity, give new impetus to the arms race at both the nuclear and conventional levels and might place greater emphasis on internal repression at home and on Soviet control of Eastern Europe. In Soviet minds the possibility cannot but be disquieting, particularly since the other major power in the East, Japan, has recently taken a much tougher stand towards Moscow and has ignored Moscow's warnings and signed a treaty of friendship with China. In Afghanistan Soviet leaders have obviously miscalulated the cost of their intervention and now find themselves very much on the defensive in warding off the verbal assaults, not merely of the West but of virtually the whole of the Islamic world. The wave of religious fundamentalism in Iran could also spread into the Soviet Muslim constituent republics which, in terms of population trends (and to some extent economic growth rates), are likely to have an increasing influence in Soviet affairs. Meanwhile, in Eastern Europe, which they regard as an essential military and political buffer against the West, the claims to legitimacy of a ruling Communist Party have again been called into question in Poland and, despite Soviet caution over Poland, the precarious nature of the Soviet hold over many of its East European neighbours has once again been demonstrated.

As Marxist-Leninists, Soviet leaders should believe that they are part of the 'wave of the future'; nevertheless

> their schooling in dialectics combined with their understanding of Russia's historical experience both before and since the revolution often leads them to a more pessimistic analysis of current trends than perhaps the facts warrant. For the dialectic principle with its attendant notions of contradiction, struggle and conflict merely tend to heighten the sense of insecurity in a country all too frequently invaded and embargoed and with a tally of twenty million dead following Germany's unprovoked and undeclared onslaught in 1941.[12]

Nor are Soviet military leaders probably very different from their Western counterparts in basing their military contingency planning on 'the worst possible case' in the need to establish what they regard as a reassuring 'correlation of forces'.

The Ideological Dimension

Yet it is often the ideological element in Soviet policies which is stressed by its critics as providing one of the main impulses behind Soviet expansionism and one of the main threats to the Western world. The Soviet Union is portrayed as not only an expansionist world power but as being ruled by a revolutionary regime seeking to convert the people of the world to the proletarian cause by sowing dissension amongst the capitalist imperialist camp and the 'peace-loving, toiling masses' of the Third World, and by building up the strength of the Commonwealth of Socialist States and in particular of the Soviet Union to the point where it may become a more effective instrument for refashioning the world in the image of the proletariat. Certainly, the idea of struggle is integral to the communist ideology; as far back as 1960 the Conference of Communist Parties stated that: 'Peaceful coexistence of countries with differing social systems does not mean conciliation of the socialist and bourgeois ideologies. On the contrary, it implies intensification of the struggle of the working class, of all the communist parties, for the triumph of socialist ideas.'[13] The Soviet leaders' secretiveness and suspicions of the outside world may reflect their ideological preconceptions of the latter's implacable hostility. It is also evident that the socialist ideology not only provides a veneer of legitimacy for defending socialist solidarity but it also seems to induce in Soviet minds a sense of rectitude when pursuing what others regard as distinctly imperialistic *realpolitik* behaviour. Nevertheless the view taken here is that ideological differences are not of the first importance, and that few conflicts spring *directly* from competing ideologies; that ideology tends rather to exacerbate conflicts which have other causes by influencing them with the heat of moral indignation; and that there is a good deal of truth in Dr Brzesinski's dictum that 'the Soviet Union's ideological appeal has flagged at home and abroad' and that the invocation of its 'ritualistic and convoluted terminology' for the most part obscures rather than clarifies Soviet motives.[14] The weight of the evidence suggests that Soviet diplomatic practice has increasingly conformed to the *realpolitik* of a great power ready to exploit opportunities of expansion but reluctantly aware of the constraints of coexistence in a nuclear world.

Nor is the ideological dimension confined to the Soviet Union. In a not altogether dissimilar fashion the Christian Church lays claim to a positive duty in terms of its own beliefs to condemn not only terror or starvation or exploitation but also the evils of the arms race and the political arrangements that allow these evils to persist. But the resurgence in the West of the view so prevalent in the 1950s of the East-West conflict as exclusively a conflict of good with evil is a disquieting reflection of a belief current amongst many Christians, particularly so-called 'born again' Christians or the Christian New Right in the United States,[15] that conflict with the Russians 'resembles that between the archangel Michael and Lucifer rather than between Tweedledum and Tweedledee'.[16] One of the dangers of this Manichaean stance, as Michael Howard has also pointed out, is that the adversary becomes dehumanised. 'He ceased to be a party with fears, perceptions, interests, difficulties of his own; one with whom rational discussion and compromise was possible.'[17] As Donald MacKinnon has well said, 'the naked blasphemy of Constantine's slogan "in hoc signo, vinces" (in this sign (that is, of the Cross) conquer) remains an outrage as disastrous in consequences as it is spiritually repulsive in underlying inspiration'.[18]

II

The Necessity of Détente

Détente is a highly ambiguous concept and needs to be used with care.[19] Here it is used in the sense of the relaxation or alleviation of tensions between the major powers, particularly between the superpowers. Détente re-emerged in the late 1960s mainly out of the nuclear adversary/partnership relationship of the superpowers. In the European context it reflected something of a compromise by the leaders of the two Germanies between their acceptance of the existing division between their two countries as essential to stability in Europe and the attempt to mitigate the injustices that division would perpetuate. In this latter respect détente has certainly achieved a good deal of success in terms of the much greater ease (despite recent restrictions) with which people can travel between the two Germanies and in the permission given to over 350,000 to be allowed to come to West Germany under the Family Reunion Programme. Nevertheless most of the official hopes of détente have been belied, particularly the hope that détente might lead to an easing of world tensions generally and lead to the more effective management of what was expected to remain predominantly an adversary relationship between the superpowers. ' "Détente" has not brought about that ever-widening co-operation ('reciprocal and comprehensive' are Helmut Sonnenfeldt's adjectives) which had been hopefully looked for since the early days of President Johnson's "bridgebuilding" '[20] The prospects, held out by the Conference on Security and Co-operation in Europe in 1975, of a thickening web of mutually advantageous links — economic, scientific, cultural — which might further help to ease tension in Europe certainly look, at the time of writing (July 1981), distinctly jaded, while military confidence-building measures have so far proved of limited practical value and arms control negotiations appear to have reached an impasse.

Indeed many critics, especially in the United States, see the whole idea of détente as an illusion and a trap for the West on the grounds that the Russians have used it as a tactical means of 'tranquillising' the West while they steadily increase their global military power. Professor V. Kortunov did in fact argue in the May 1979 issue of *International Affairs* (Moscow) that 'for the Soviet Union and other Socialist states, the policy of détente constitutes a struggle to create favourable conditions for Communist construction and for the development of the world revolutionary process'.[21] Moreover, in Soviet eyes détente (or, as they prefer to call it, 'peaceful coexistence') is evidently divisible, since the Soviet Union is in terms of its ideology expected to support so-called 'just' wars of national liberation and to foment unrest within the capitalist world generally. Consequently, détente to the Soviet Union would seem to be limited to the continued recognition by both superpowers of their joint interest in avoiding nuclear war, in providing some kind of brake on the nuclear arms race, and in accepting the legitimacy of existing European frontiers (including the anomalous, and still vulnerable, position of West Berlin). Yet even this limited notion of détente, of controlling tensions which might threaten stability in Europe (the most likely arena of nuclear war), is not to be derided. The concept of the Harmel Report of 1967 that 'military security and a policy of détente are not contradictory but complementary' still remains valid. As Dr Hans Apel, the West German Federal Minister of Defence, recently put it

> Defence capability and readiness to pursue détente must serve to consolidate security in Europe, broaden the basis of cooperation between East and West, and thus establish the conditions precedent to limiting — or at least de-fusing — existing issues of conflict despite political, economic and ideological antagonisms.[22]

In the longer term, if international society is to survive at more than a minimal level of order and the more dangerous international tensions are to be successfully contained, the major powers, and particularly the superpowers, must learn to exercise their power with restraint, to acknowledge each other's essential security interests and to act together to check conflicts which threaten to escalate to global proportions. None of this is easy. The element of rivalry between the superpowers is likely to outweigh that of common interest in all but a very few isolated cases. Perceptions of security interests themselves are apt to be highly subjective — and fluid, as Suez, Vietnam and Afghanistan confirm. And there is always the risk of 'adventurism', miscalculations and misperceptions and the endless complications of prestige, domestic electoral worries, leadership struggles and so on, as well as the possibility of rash acts by lesser powers which can put the superpowers in extremely difficult situations not of their own mking. Nevertheless all this makes it more rather than less necessary for the superpowers to achieve as clear an understanding as possible of the 'ground rules' of behaviour (particularly the need for extreme caution in threatening or using force) and to secure improved techniques of crisis management, so that they are more likely to draw back from the brink whenever a crisis threatens to jeopardise their common interest in avoiding a nuclear holocaust.

> Détente in the 1980s, even more than in the previous decade, will scarcely be able to solve any of the major outstanding problems be-

tween East and West. But perhaps it can reach the modest objective of helping to manage the crises that are bound to occur. Even this will require not less but more dialogue between East and West, particularly between the two major powers.[23]

Yet for Christians détente must mean much more than that.

As Christians immerse themselves in the totality of the Christian tradition, seeking to discern the nature and will of a God who creates, judges and redeems us, one central theological motif which has profound ethical implications to the Christian life is 'reconciliation'. The work of Jesus Christ can be understood in a fundamental way as a work of reconciliation — 'God was in Christ reconciling the world to himself' — and the Christian life in response to that work is to be 'agents of reconciliation'. To the eyes of faith there is a unity to humankind, a given oneness that is belied by our visible antagonism and alienation. A central conviction of our faith is that the forces serving reconciliation in community are to be affirmed over those serving division and enmity.[24]

This is not to lose sight of the fundamental differences between the relatively 'open' societies of the West and the more 'closed' societies of the Soviet system — it was not the West that built the Berlin wall. Nor is it to disregard the realities of power and the constraints under which political leaders on both sides operate. Nevertheless it is to stress the Christian duty to keep alive the hope of mutual antagonisms and tensions being diminished to the point where genuine, deep and mutually agreed reductions in the present appalling size of national armaments, particularly nuclear armaments, becomes possible.

Arms Control and Disarmament

It is unhappily the case that the strains on détente, the concern over the build-up of Soviet military strength in Europe, particularly in medium-range strategic nuclear systems, and the considerable Soviet capability for offensive chemical warfare, all combine to make the climate for arms control initiatives distinctly unpromising.[25] Even arms control agreements based upon equivalence of concessions whether in theatre nuclear weapons or in conventional armaments would in present circumstances, it is claimed, perpetuate — even deepen — present Western inferiority in the European theatre.

A fundamental difficulty is that much of deterrence thinking still portrays a rigidity of outlook which is a serious obstacle to meaningful arms control negotiations. It is true that

> The outstanding importance of mutual perceptions, or, rather, the need to learn more about the adversary is acknowledged; so also is the need for continuing communication and the political consistency that, more than mere military superiority, endows deterrence with its essential credibility.[26]

Nevertheless

> The official consecration of strategic parity never fully eradicated the United States' innermost conviction about her superiority and the need to keep it.[27]

Moreover, there is the powerful argument that the Soviet Union can never accept a

> mutuality of deterrence as long as she, unlike the United States, is confronted with more than just one major military adversary. It has not therefore simply been a traditional preference for over-insurance in large numbers that has led the Soviet Union to stress so firmly the military dimensions of her policy, but rather a political environment fundamentally different from that of the United States.[28]

For reasons such as these many Europeans are inclined, as Michael Howard points out, to take a less alarmist view than most Americans of the Soviet arms build-up and to see it as reactive rather than as a deliberate bid for world domination.[29] He and others (for example McGeorge Bundy)[30] have also cast some doubt, on military grounds, on the need for cruise missiles and Pershing IIs to act as a counterweight to Soviet SS.20 missiles and Backfire bombers. The ageing of NATO's medium-range nuclear systems may make some measure of NATO theatre nuclear modernisation sensible; and in any case to forgo that option when just on the point of entering negotiations to limit medium-range nuclear systems on both sides would remove much of the incentive for the Soviet Union to negotiate seriously. But it is still essential to match the modernisation programme with a genuine and determined intent to negotiate, as indeed the NATO directive of 12 December 1979 stressed.

What must never be forgotten is that a security system which seeks to maintain a balance through ever-increasing and ever more costly nuclear armaments is potentially disastrous. As each side tends to take a pessimistic view of the other side's capabilities and intentions — as they habitually do — each side sees the need to catch up with the other and attain a new balance at a higher rather than a lower level. More than once in the past fears of nuclear inferiority (for example the so-called missile 'gap' in the late 1950s) have proved a myth and have merely provoked the other side to strive to close the gap which *they* then perceived. Moreover, although the awesome potential of certain new types of weapons has been recognised by the scientific community which has sometimes been instrumental in securing their limitation by treaty (for example the Peaceful Use of Outer Space Treaty and the Anti-Ballistic Missile Treaty), only too frequently much of the impetus behind the arms race on both sides comes from the 'man in the laboratory' intent on outwitting his opposite number and from a military-industrial complex with a deep, long-term commercial and psychological involvement in the race.[31]

This kind of leap-frogging deterrence is a recipe for spiralling madness. It also constitutes a wasteful diversion of wealth within the domestic economies of states, ties up human and material resources which *might* otherwise help to reduce Third World poverty and brings the possibility of a nuclear exchange considerably closer. Although recent developments in strategic doctrines (for example Presidential Directive 59) are designed primarily to make deterrence more credible by updating the long-standing NATO doctrine of 'flexible response', the 'greater attention' being given to 'how a nuclear war would actually be fought by both sides if deterrence fails'[32] runs the risk of encouraging the illusion that the controlled and limited use of nuclear weapons for battlefield missions, let alone for theatre missions or a selective counter-force strike, could be prevented from escalating rapidly to a full-scale nuclear war. In particular, the argument that the West cannot agree to an explicit renunciation of the first use of nuclear weapons, on the grounds that its conventional forces cannot by themselves be relied upon to serve as an effective deterrent to a full-scale Soviet conventional attack, is very difficult to justify.

Outside Europe the possession of nuclear weapons is only indirectly relevant to the task of combating Soviet expansionism and there is the danger that if the West were to maintain too low a level of conventional forces, it might be driven on grounds of prudence to back down in crises in 'grey' areas since not to do so could lead to a war which could not be 'won' without resorting to nuclear weapons.[33] All this underlines the danger of NATO concentrating unduly on modernising nuclear forces or of Britain perpetuating largely irrelevant national nuclear capabilities (for example the UK Trident) at the cost of paring down conventional forces which are already under considerable strain to meet the tasks remitted to them.

Thus, perceptions both at the domestic level and within NATO of the dangers of a spiralling arms race which could accentuate the already appalling cost and aggravate the risks of nuclear war are neither without foundation nor are they a political factor which can be ignored. Ultimately the security of the West depends upon the good health of the Western polities and on the willingness to make the necessary domestic sacrifices to secure the external defences. Moreover, there is a risk of technological advance undermining the concept of a stable balance of nuclear deterrence between East and West and opening up the possibility of one side attaining a first-strike capability (or of being thought by the other side to have attained, or to be about to attain, such a capability, which is just as dangerous). Given the number and variety of strategic weapon systems in existence the acquisition of a crippling first-strike capability may seem highly improbable. But it is subjective perceptions of capabilities and of changes in them which often determine policy irrespective of objective realities.

Consequently, although the international climate for meaningful arms control negotiations may be discouraging, a determined effort needs to be made to achieve greater security at a lower level of nuclear confrontation. A first step to minimising technological competition and the mutual suspicions and anxieties to which it gives rise would be for each side to consider more carefully the likely impact on the other of embarking on new weapons programmes. Above all, there is the need to keep negotiations going which 'might improve mutual knowledge and understanding of capabilities, plans and doctrines, thereby enabling priorities to be determined with greater care, and in a manner not unduly provocative to the other side'.[34] This does not mean large set-piece negotiations with a remit to produce comprehensive agreements, but more limited exchanges developed in response to specific problems. 'The most radical practical step that is worth considering is the prohibition of battlefield nuclear weapons from forward positions in Central Europe.'[35] Then there is the French proposal of December 1980, backed by NATO, for a meeting to negotiate new confidence-building measures which would be militarily significant, verifiable and binding and applicable to all of Europe, including the European part of the Soviet Union, and which might help reduce the fear of surprise or pre-emptive attacks. There is also a need for measures to counter the acquisition of anti-satellite capabilities which could, *inter alia*, threaten existing systems of communication, command, control, targeting and surveillance; and generally measures which would minimise the risks of destabilising technological breakthroughs at the nuclear level and conduce to a measure of accord on the nature of the strategic environment. At the very least negotiations can, as in the case of SALT I and II 'serve as a framework for the avoidance of major miscalculations and misunderstandings, and . . . provide a boundary for strategic programmes and thus a degree of predictability in the strategic competition'.[36]

Lastly, there is the need for the preservation and tightening of the restraints on the proliferation of nuclear weapons, particularly through the co-operation of suppliers and a strengthened International Atomic Energy Agency, and in such a way as to protect the legitimate energy needs of the developing countries and their interest in the peaceful uses of nuclear power. This is dealt with more fully by Roy Dean in Chapter 6. Here it is only necessary to stress two points. The first is that there is a minimal level beyond which reductions in the nuclear arsenals of the superpowers should not go, lest their capacity to take effective collective action to check or limit nuclear war between states with newly acquired nuclear capabilities be called into question. The second is simply to underline the dangers of unchecked proliferation not only for local contestants but for the superpowers as well.

> Accelerated proliferation, especially among 'enemy pairs' of states (India/Pakistan, Israel/Iraq, Argentina/Brazil, etc.), could not fail to increase both local insecurity (given both mutual vulnerabilities and underlying political tensions) and to affect adversely the superpowers' contest, since the acquisition of nuclear weapons by the enemy of a great power's client is more likely to incite that great power to shore up the client than to promote mutual dissociation by the superpowers, at least in areas of vital importance to them.[37]

III

Principle AND Prudence

In diplomacy, as in politics generally, there are many differing tasks, but in a quasi-anarchical international society one is bound to stand out above all others, that is the safety and security of the state. Those who wield, or are under, authority in the state bear corresponding responsibilities. Those who are Christians also have to act in an increasingly secular society and to work alongside colleagues many of whom are not Christians. They will be only too well aware of the constraints that so often allow only of an approximation to the ideals of the Christian ethic; that the principles to be derived from that ethic frequently conflict; and that politics is indeed the 'area where conscience and power meet, where the ethical and coercive factors of human life inter-penetrate and work out their tentative and uneasy compromises'.[38]

All that most Christians will probably feel called upon to do is to pray that their leaders, both secular and spiritual, will have a 'right judgement' in their pronouncements and their policies. Others, however, will feel impelled to recall their fellow Christians to the lessons of the Cross and to what they regard as the sinfulness of relying upon the deterrent power of nuclear weapons even for self-defence (as Bruce Kent argues in Chapter 3). Their more prophetic voice deserves respect — and attention. But the point I want to make here is that there is a major responsibility on Christian churches to face up to the dilemmas posed, more acutely perhaps in a nuclear age than ever before, by the often competing claims of peace, order and justice and to the complexities of the management of power in international society. A responsible attitude is not helped by indulging in idealistic prescriptions which ignore the realities of power in favour of simplistic notions of international right and wrong. What is needed is a vision of the ideal, of the reconciling act of God in Christ and that the world is indeed one in the sense that it is 'God's world', coupled with a grasp of the parameters of the possible; an awareness of the wonder of God's redemptive love and of the workings of Divine Providence in history with a realisation of the necessarily fallen character of a quasi-anarchical international society. In short, what is needed is a concern not only with what states *ought* to do, but also for what they *can* do.

Above all, conflict in international society needs to be seen, not so much in terms of the conflict of good with evil (though at times it is that), but more as an inescapable part of the human predicament in a society lacking any overall 'common authority to keep them all in awe' (Hobbes). Herbert Butterfield, a great historian and Christian, reminds us that the tragic element in international conflict is that such conflict is not a simple picture of good men fighting bad; rather we need to appreciate that the very structure of international society produces situations of Hobbesian fear. The tragedy is that conflict is so easily 'embittered by the heat of moral indignation on both sides, just because each is so conscious of its own rectitude, so enraged with the other for leaving it without any alternative to war'.[39] It is this conflict between 'embattled systems of self-righteousness', in which men delude themselves into thinking they are gods and that 'inhuman means are justified by the superhuman ends'[40] which is so disruptive of our shared humanity. Cosmic humility is a Christian virtue.

Finally, in all these matters principle needs to be married to prudence. In political choices principles must never be lost to sight, but their affirmation and application needs to be accompanied by a prudential calculation of the political consequences to which they are likely to lead. 'In classical and Christian ethics the first of the moral virtues is *sophia* or *prudentia*, because without adequate understanding of the structure of humanity, including the *conditio humana*, moral action with rational co-ordination of means and ends is hardly possible.'[41] In a nuclear world prudence is still a Christian virtue — and a recipe for survival.

Notes

1. See p. 75 below.
2. See p. 103 below.
3. H. Morgenthau, *Dilemmas of Politics* (University of Chicago Press, Chicago, Illinois 1958), p. 83.
4. Article 2 (4) reads: 'All Members shall refrain in their international relations from the threat or use of force against the territorial integrity or political independence of any State, or in any other manner inconsistent with the Purposes of the United Nations.'
5. *Strategic Survey, 1980-1981* (International Institute for Strategic Studies, London, 1981), p. 14.
6. Ibid., p. 15.
7. Geoffrey Stern, 'The Soviet Union, Afghanistan and East-West Relations', *Millennium* (Summer 1981). See also House of Commons, *Fifth Report from the Foreign Affairs Committee*, Session 1979-80 (HMSO, London, 30 July 1980), p. ix.
8. Stanley Hoffman, 'Security in an Age of Turbulence: Means of Response', in *Third World Conflict and International Security*, Part II, Adelphi Paper no. 167 (International Institute for Strategic Studies, London, 1981), p. 13.
9. Seweryn Bialer, *Stalin's Successors* (Cambridge University Press, Cambridge, 1980), quoted in Stanley Hoffman, 'Security in an Age of Turbulence', p. 3.
10. See memorandum by James Mayall, 'The Soviet Union in Africa', in *Fifth Report from the Foreign Affairs Committee*, pp. 161-5.
11. See Michael Binyon's assessment, *The Times*, 22 October 1980.
12. Geoffrey Stern, 'The Soviet Union, Afghanistan and East-West Relations', p. 5.
13. 'Statement of the Conference of Eighty-one Communist and Workers' Parties, Moscow' (6 December 1960), in *Documents on International Affairs, 1960* (Oxford University Press for Royal Institute of International Affairs, London, 1964), p. 229.
14. Quoted in Elizabeth and Wayland Young, 'Marxism-Leninism and Arms Control', *Arms Control* (May 1980).
15. Peter Berger, 'The Class Struggle in American Religion', *The Christian Century* (25 February 1981).

16. Michael Howard, *Studies in War and Peace* (Temple Smith, London, 1970), p. 171.
17. Michael Howard, *War and the Liberal Conscience* (Temple Smith, London, 1978), p. 128.
18. Donald M. MacKinnon, 'Power, Politics and Religious Faith', *British Journal of International Studies* (April 1980).
19. See Brian White, 'The Concept of Détente', *Review of International Studies* (July 1981).
20. Elizabeth and Wayland Young, 'Marxism-Leninism and Arms Control', p. 7.
21. Ibid., p. 11.
22. In a paper presented to the Wehrkunde Conference (1981), p. 6. See also Richard Lowenthal, 'The Shattered Balance', *Encounter* (November 1980).
23. *Strategic Survey, 1980-1981*, p. 8.
24. Allan M. Parrant, 'The Sermon on the Mount; A Theology of Reconciliation and International Politics', paper presented to the international conference of the Council on Christian Approaches to Defence and Disarmament, Friedewald, West Germany, 1980.
25. The IISS (*Strategic Survey 1980-1981*, pp. 6-7) takes a less pessimistic view and argues that 'the Soviet Union might be willing to be more co-operative than hitherto' and that it would 'seem wrong to regard the Soviet Union's proposals for new arms control as mere propaganda'.
26. Curt Gasteyger, 'The Determinants of Change: Deterrence and the Political Environment', in *The Future of Strategic Deterrence*, Part II, Adelphi Paper no. 161 (International Institute for Strategic Studies, London, 1980), p. 7.
27. Ibid., p. 8.
28. Ibid., p. 7.
29. Michael Howard, 'Return to the Cold War', *Foreign Affairs* (Spring 1981).
30. McGeorge Bundy, 'Strategic Deterrence Thirty Years Later: What Has Changed?', in *The Future of Strategic Deterrence*, Part I, Adelphi Paper no. 160 (International Institute for Strategic Studies, London, 1980), p. 10.
31. See Lord Zuckerman, *Science Advisers, Scientific Advisers and Nuclear Weapons* (Menard Press, London, 1980), pp. 10-11.
32. Harold Brown, then US Secretary of Defense, *The Objective of US Strategic Forces*: An Address to the Naval War College in Washington, 20 August 1980 (International Communication Agency, US Embassy, London, 1980). But see statement by then Secretary of State, Edmund S. Muskie, before the Senate Foreign Relations Committee, 16 September 1980, in which he stresses that PD.59 'does not signify a shift to a war-fighting strategy nor to a first-strike doctrine' (US Department of State, Current Policy no. 219, September 1980). Also Barrie Paskins in Chapter 5, pp. 95-6 below.
33. Moreover, as Michael Donelan remarked to the group, if backing down became a habit, it could eventually lead to an attitude of stubbornness — 'so far and no further' — and to a decision to adopt a nuclear first-use policy.
34. Lawrence Freedman, *Arms Control in Europe*, Chatham House Papers, no. 11 (The Royal Institute of International Affairs, London, 1981), p. 46.
35. Ibid.
36. Christoph Bertram, *The Future of Strategic Deterrence*, Part I, Adelphi Paper no. 160 (International Institute for Strategic Studies, London, 1980). p. 3.
37. Stanley Hoffman, 'Security in an Age of Turbulence', p. 15.
38. R. Niebuhr, *Moral Man and Immoral Society* (Charles Scribner's Sons, New York and London, 1932), p. 4.
39. H. Butterfield, *History and Human Relations* (Collins, London, 1952), p. 21.

40. Eric Voeglin, *The New Science of Politics* (University of Chicago Press, Chicago, Illinois, 1962), p. 169.
41. Ibid., p. 165.

2 THEOLOGICAL METHOD IN THE DETERRENCE DEBATE

G.R. Dunstan

Christian thinking on war and peace, defence and security, over the last thirty years has been marked increasingly by its respect for the facts of the case, its insistence on giving full weight to the empirical features as responsible experts in the relevant fields present them. Christians, including theologians and moralists, have sat with professional diplomatists, military men and political commentators — themselves also Christians — studied their analysis, and sought out the points of moral claim. They have then searched their theological and moral traditions for an apt wisdom. The regular response of ecclesiastical bodies to reports produced in this way is that they are 'not Christian enough'; 'lacking in theological content'; 'they might have been produced by any secular body'. We have to consider what such commentators expect, and whether their expectations can be met.

If they assume that there *is* a specific 'Christian approach', perhaps even specific 'Christian' solutions, to problems of security, that assumption must be questioned. If the language and meaning of Christianity are taken seriously, there are some human activities which cannot be discussed in Christian terms at all. There is no specifically 'Christian' way of waging war, or of amputating limbs, or of fixing oil prices, or of deciding for or against the nuclear generation of energy. (The personal motive and conduct of a Christian soldier, or surgeon, or oil magnate, or nuclear physicist may well be influenced by his Christianity and Christian character; but that is a different matter. The matter here is the nature of activities.)

The statement of the Lambeth Conferences of 1930, 1948, 1968 and 1978, that 'war as a method of settling international disputes is incompatible with the teaching and example of Our Lord Jesus' is a truism that cannot be gainsaid; there is no 'Christian' way of prosecuting an inherently unchristian pursuit. (With making peace, before or after a war, it is otherwise; there is a Christian language for this, and a Christian reality, in forgiveness, reconciliation, new life through sacrifice, rebirth into community or fellowship.) What passes too often for 'a Christian contribution' to world problem-solving is readily exposed for what it is: a veneer of Christian, or perhaps biblical, language upon policies dictated by other considerations — a judgement fairly made upon some of the products of the World Council of Churches, and on some 'liberation theology', Roman Catholic as well as other. That Christians engage in these activities, and have a duty to do so, is not denied; but the categories which they employ and the solutions to which they come are not Christian categories or solutions, for there are none specific to these activities. The centuries of work in the traditions of natural law and the 'just war' point the same way: canonists and moralists had to go outside the strictly theological tradition, into those of philosophy and law, in order to work on the world's problems to any effect at all. The 'just war' tradition is not in origin or in itself a Christian tradition; Christian thinkers took it over from pagan Greece and Rome, adopted it, brought it up, developed it, and gave it back, adult, to the modern world. When Grotius modulated *iustus et pius* into *iustus et legalis* he did not change the substance of the doctrine of the just war, but only the language: Christians had been operating an essentially human, that is secular, discipline, which was now coming back into its own.

If it is assumed, further, that a 'Christian contribution' must be 'biblically based', we are driven to enquire what is the relevance of Holy Scripture to our search for security, and how should Holy Scripture be interpreted and employed. If theologians are invoked, they must insist upon an exegetical and theological handling of the Scriptures which asserts both a radical continuity and a radical discontinuity between the Old Testament and the New — and both highly relevant to the task in hand.

The Old Testament is the literature of a political community preoccupied, in every century, with its own security — as a group of nomadic tribes; as settlers in Egypt; as conquerors in Canaan; as kingdoms bestraddling the trade and military routes between warring empires, and so subject to invasion, siege, defeat, destruction; as captives in Babylon; as returning exiles enjoying brief centuries of precarious liberty until conquered again by the Seleucids and the Romans; as rebels who, in AD 135, were crushed and dispersed. Within this history they gained both religious and political experience. Their religious experience was of the God who revealed himself to them, interpreted by Moses, the prophets, the psalmists and the wise; and in this experience of God they came to a responsive judgement upon themselves, upon good and evil, right and wrong, blessing and curse in man. Their political experience was related to their religious experience, as both were shaped by their leaders. The prescriptions for security were *political* prescriptions: invade this territory, respect the frontiers of another; go out to battle, refrain from battle; ally yourself with this nation, do not become entangled with another; fortify your city, withstand the siege at one time — surrender and save yourselves at another. That was the hard political advice, given out of political and military experience. It was given also out of religious conviction, expressed in unremitting calls for *faithfulness*, an active fidelity to YAHWEH their own God — indeed, the true God — and so for a refusal of alliances with Egypt and other nations professing other gods: a concept to be distinguished from a passive 'faith' in God which would leave the issue to him — a 'religious' solution — without human political or military activity at all. Within this religious, political and military experience were realised and worked out the specific religious insights of the Jews: the righteousness and holiness of God; the fused sublimity and sinfulness of man; the reality of vicarious suffering; the need of expiation, forgiveness, redemption, restoration; and the hope for a redeemer, one anointed with God's spirit, who would inaugurate God's rule or kingdom — though whether as political messiah or a son of man gloriously transcending all thrones and dominion, expectations were divided.

There is a radical continuity of the New Testament with these Old Testament Scriptures in that the earliest Christians, seeking a language in which to interpret their experience of Jesus and their deepest convictions about him, found it in this language of expectation, and so clothed him with it. (The extent to which he gave them clues to this interpretation is a study beyond the present task.) There is a radical discontinuity of the New Testament with the Old in that, whereas, like the Old, the New Testament was the product of a community, that community never saw itself as a political community nor acted as one. It was its very determination to embrace the universalistic as against the particularistic strand in Judaism — to break down the wall of partition between Jew and Gentile — which brought upon it the hostility first of political Jewry, then of political Rome. That it was a non-political community, transcending (without obliterating) all natural, political and national distinctions, united only by a common faith in Jesus as the Risen Lord and a common possession of his messianic Spirit, is evidenced indisputably by the story of Pentecost in Acts, by St Paul's pregnant utterances in Galatians 3:28[1] and Colossians 3:11,[2] and by the whole tenor of the Epistle to the Ephesians. The politics of national survival were irrelevant to it. Jesus, on all the evidence available, carefully dissociated himself and his mission from that of a political messiah, and from the anti-Roman revolutionary groups active all around him in his day. There is no evidence at all that the earliest Christian communities took political action to implement their theological transcendence of imposed distinction, as between bond and free. In all the interpretations which they left us of the death of Jesus on the Cross, there is not one hint of a promise attached to it of political success, or of its use, actual or potential, as a political weapon: they thought of it entirely within the purely religious idiom of the various traditions of sacrifice, whether of animals or of the righteous servant of God.

In relation to the world, its politics and its peace, the Christian communities — or such of them as have left us evidence — were stretched between two convictions. The strongest — which Christians today, whatever they may profess, certainly do not live by — was that the Lord would come again in their lifetime and establish that rule or Kingdom of God which He had already won by His death and resurrection. What are now called the ethics of the New Testament are dominated by this expectation, this hope. Its practical application dominates all St Paul's moral counsels — as in 1 Corinthians 7; its theological force is at its strongest in Romans 8:18ff., where it is set over against 'the sufferings of this present time' under which the whole creation 'groaned', and Christians with it. There is no hint that these sufferings are to be fought against: all the emphasis is on expectant hope, that we should soon be delivered from them by God's hand. Such as could be relieved by Christian charity were to be relieved — that is evident everywhere. But of political action to relieve them there is no hint. The other conviction, however, was that politics was a valid human activity, and, indeed, God's work through men. Part of the moral exhortation drawn (habitually) by St Paul out of his theological exposition in Romans enjoins 'subjection to the higher powers' for 'the powers that be are ordained of God', ministering, from God, good to those who do good, and evil to those who do evil. Taxes and tolls are to be paid to them on this account, to support their work — work from which St Paul profited (as St Luke, his biographer, was careful to emphasise) in the protection given him by the Roman magistracy and the Roman power. In a later epistle, I Timothy, Christians are exhorted to offer intercessions and thanksgivings 'for kings and all that are in high place, that we may lead a tranquil and quiet life in all godliness and gravity'. The homily enshrined in 1 Peter parallels the Pauline teaching on the subjection due to kings and governors in a passage which ends (2:17) by putting the religious duties of Christians firmly within their political duties: 'Honour all men. Love the brotherhood [that is the Church]. Fear God.

Honour the King.'

The religion of the early Church, therefore, was neither 'other-wordly' nor 'this-worldly' – but 'both-worldly'. While the world lasted it had to be governed. It was God's world, and secular rulers ruled by his ordinance, with a derived authority; and the sword was not borne in vain. Within this world order, and under the protection of the civil power, the earliest Christians tried to live as in God's Kingdom, as though God's rule had actually begun. They were genuinely surprised when after baptism members of their communities fell into sin; genuinely surprised when some died before the 'end' came. They made no political attempt to change the world – to abolish slavery, or to challenge an empire resting on military might. When the Roman emperors persecuted, they denounced them in terms of biblical and inter-testamental apocalyptic (as in Revelation); but they did not renounce their faith in civil dominion as such: they continued, as the liturgies show, to uphold civil magistracy in their common eucharistic prayer. They refused themselves to undertake magistracy, or military service, or any office under the emperor, because of the idolatrous oath required upon entry on office. In their own defence they claimed that they served the empire well enough – and better than other men – by the efficacy of their prayer, and by their peaceful, law-abiding, socially benevolent way of life.

If, now, a 'specifically Christian' element be asked for in the contemporary search for security, or if a 'biblical perspective', to which of these elements do expectations point? From which of them can the expectation be met? Certainly not, if exegetical and theological integrity be prized, by once again treating the tribal bellicosities and political prescription of the Old Testament as divine commands and so justifying holy wars, crusades and the denial of humanity to the barbarian or the infidel – an abuse which medieval and renaissance canonists sought to restrain by appeal to the 'just war' tradition.[3] Certainly not, also by reading off political prescriptions from the command to love (which is as clear in the Old Testament as in the New); or from such words of Jesus as 'Put up thy sword into its sheath . . . ', or 'Peace I leave with you, my peace I give unto you'; or even from the language and example of sacrifice. All of these, and more also, are ingredients in a Christian ethics; but even a Christian ethics in its entirety cannot be transcribed directly, and without complement, into politics; and to quote such texts as though they could determine the issues of war and peace is to mistake the nature of Scripture, ethics and politics alike.

When the policy of the emperors changed, and Christians became first tolerated and then actively encouraged to take part in maintaining the fabric of the empire, their thinkers, notably St Ambrose and St Augustine, began to formulate an ethic, not simply of Christian participation in politics, including military service, but of politics and war itself. And in order to do so, they brought, quite properly, into Christian thinking categories and norms already developed outside it, in Greek philosophy and Roman law, notably the already formed 'just war' tradition; and they could do so because St Paul had already validated the appeal of the 'natural law' to the Gentile conscience in Romans 1 and 2. (Here St Paul asserts that the Gentiles, to whom God had not revealed his nature and will in terms of the Law, or Torah, as he had to the Jews, could nevertheless perceive both by natural reason 'through the things that are made', that is through the natural order. These few verses were to become a bridge between philosophy and theology of incalculable importance for the development of Christian and Western thought.) There was still no identification between the political realm, the domain of Caesar, and the spiritual realm, God's Kingdom and His Church, as St Augustine's *City of God* made clear. But Christians had now, for the first time, to work out their Christian obedience in terms of a responsibility for this world, the political realm, and its security and order, as they had not before. The one specific 'Christian' contribution which St Augustine added to the 'just war' tradition was that Christian soldiers must maintain purity of motive and love in their hearts in every act of their military service. From that time on, the 'specifically Christian' contribution to the politics of security has been the application of Christian minds to the business of politics, *working within the terms and categories given them by politics itself*, and seeking to humanise the practice, in ends and in means, adding authority to their judgements by invoking the righteousness of God as proclaimed chiefly by the Old Testament prophets, and backing them, when possible and necessary, with the weapons of ecclesiastical power, notably excommunication and interdict.[4]

It would be artificial to seek to restrict the Christian contribution to those terms now. Many new strands have developed in the tradition, new initiatives have developed. Christians recognise their duty, not only to participate in politics, but also to raise the sights, the goals of politics. It is noteworthy, however, that when they have taken such initiatives in the past — for instance for the abolition of slavery, for penal reform, for the protection and advancement of the labouring classes, for seeking humane restraints on the growing technologies of war — they have used the instruments of politics, legislative and consultative processes, to make their initiatives effective. Contemporary Christian statesmanship looks the same way. Recent magisterial Roman Catholic social teaching — as in *Pacem in Terris* (1963), *Gaudium et Spes*, the Pastoral Constitution of Vatican II (1965), *Populorum Progressio* (1967) and *Octogesima Adveniens* (1971) — constantly calls the Christian faithful to the loftiest ideals of peace, social benevolence and ideological concord; but the appeal is consistently to natural justice, perceived and morally assented to by natural reason, in the light of the Christian Gospel; and the imperative is to achieve these ends by constitutional means, by involvement in the processes of secular politics. There is a prophecy of apocalyptic despair, calling the faithful out of a political system which, because it is so evil, must fall or be brought down, entrusting the future to the hand of God. There is also a prophecy of redemptive hope, bidding Christians commit themselves to the work of politics, precisely because of its wrestling with the sin of the world, as part of their obedience to God.

The work of Christian theologians and moralists today who, in pursuit of security, sit with professional diplomatists, military men and political commentators and explore the problems within their given terms — as they will sit in citizen debate anywhere on common civic concerns — is entirely consistent with this constitutional tradition. They recognise that theological and exegetical expertise (to say nothing of a true piety or ardent charity) are no substitute for professional competence; and that Christian commitment by no means eases the burden of responsible decision — in the advice given to Ministers of the Crown, for instance, in relation to military or foreign policy — but rather increases it. The faith of a contemporary Christian, like that of his first-century counterpart, can be neither 'this-wordly' nor 'other-worldly'; it is exercised, at full stretch, in both.

Perhaps, for the avoidance of confusion, clarification should be sought in the use of such words as 'theology', 'prophecy' and 'faith' in this discussion. Theology is an intellectual discipline — for long, indeed, 'the queen of the sciences' when men gave the Latin connotation to that word 'science'. It possesses an integrity and an autonomy of its own in that it handles a corpus or body of material of its own in a disciplined way. Its language and usages should have sufficient objectivity to be capable of reasoned analysis and discussion anywhere; for it is, in its nature, an application of reason to the things of God, primarily the self-revelation of God. Faith is primarily the acceptance of, and self-commitment to, that Divine self-revelation. Theology is then, in a classic definition, *fides quaerens intellectum*, faith seeking an understanding of itself (though there are contemporary theologians who profess no faith). Faith is prior to theology and superior to it. It fashions minds, moulds character, sustains in duty and against temptation, impels to new initiative, creates new order; it is the channel of God's grace for a new created world. Prophecy is a product of faith — though its authenticity is not to be measured solely by the degree of passion in its utterance. Prophecy is in essence a proclamation of divine truth. As such, it is powerful to move even the mountains of political intractability; though they cannot be expected to remove at the mere sound as the walls of Jericho fell before the blast of the trumpets: the word must often be incarnate in political process or diplomatic activity. Prophecy can become false prophecy, either by clothing deceptive dogmas with the language of religious passion or persuasion, or by presuming, beyond the evidence, some particular political means to be inherent in the given religious end. On the evidence of the New Testament, the duty to test the spirits is as old as the existence of Christian prophecy itself; and the test is that of reason.

It may therefore, and properly, be asked: what contribution do theologians bring to this political exercise? Their hardest duty is to maintain in full vigour the authentic *contradictions* which are firm in the tradition, scriptural and theological. They may not simplify, reduce or become partisan. They have to reconcile a belief in God's righteous providence in and for His own world with belief in man's instrumentality within larger parts of it — with a hard-won dominion, based on accumulated knowledge and skill within some parts of physical nature; with political dominion in human government and social affairs. They have a doctrine of man — that he is made in the image of God, free yet of infinite worth to God and destined for a life of fellowship with God; that this image is defaced by sin, so radically that frustration enters into all human affairs, denying even to utterly good acts the guarantee of an utterly good consequence or response; that man is nevertheless redeemed, so that an ultimate hope must exclude an ultimate despair; that the righeous God can and does lead men, by His Holy Spirit, to an awareness of and repentance for their own sins, and so to perceive the righteous and the good and to follow it; that forgiveness is a reality, freeing men and nations from bondage to the past for reconciliation and a new relationship,[5] and that political activity is as much within this grace — this gracious activity of God — as any other activity of man. To keep this balance is to walk between a facile optimism — a belief that all things are possible to us once we have perfected the system, whether it be a weapons system or an instrument for world government — and a numbing despair, the assumption that the worst must happen, so that there is nothing we can do about it. This realism about man is a preservative, also, against false polarities, dividing men and nations into angels and devils, sons of light and sons of darkness, with ourselves wholly good, wholly right, and others, the enemy or potential enemy, wholly bad, wholly wrong. (The debate between pacificist and militarist can also run out to these extremes.) Granted, too, this doctrine of man in its fullness, it is possible to seek for an ordinate balance between self-interest, self-regard, and altruism, other-regard, and this between nations as well as between individuals. Supremely, the doctrine bids us, on the one side, to strive for peace, to labour to build it and to keep it, and, on the other, to provide for the possibility that peace may fail; in short, to arm. In this good yet disordered human realm there is no security without authority, no authority not ultimately backed by power.

Security is not always, however, synonymous with justice. Tyrannies can make themselves secure; order can be imposed without justice; authority can employ power as an instrument of evil. There is paradox, contradiction, again in the tradition about this. This is not the place to develop the differences between the 'righteousness' of God in the prophetic tradition and 'justice' in the Graeco-Roman tradition; there are differences, despite the translation of 'righteousness' by the Latin word *justitia*, from which difficulties arise in the cognate words 'justify' and 'justification'. The prophets declared the God of the Hebrews to be a 'righteous' God, who 'shewed' His righteousness to Israel in establishing them as a nation and in defending them against their enemies. He called upon them, in return, to live 'righteously', that is, in practical terms, to establish what we now call 'social justice' among themselves and to be faithful to Him in their covenant. Nevertheless, Israel was not always safe from her enemies, and in political terms Israel as a nation was destroyed. In personal terms, also, the righteous did not always prosper; the wicked triumphed and the right went by default; the Psalms are eloquent upon the theme. When vindication, justification, did not come, to the nation or to godly men, in this world or age, they began to look for it in an 'age to come'; an earthly hope became an eschatological hope — God's righteousness will triumph in the End.

The contemporary world faces the same paradox in different terms. Political tyrannies, sustained by the ever-increasing powers conferred by new technologies, become ever harder to overthrow from within. The old optimistic notion, that a state founded on injustice cannot last, affords less comfort. In the nineteenth century a compound of Christian and political moralism, backed by strong military force, launched local punitive wars to vindicate the wronged, to punish massacres or gross political injustice. Today the process is seen in reverse. Revolutionary Communism is actively engaged all over the world, in fostering, by armed internal revolt or even by direct invasion, 'liberation' from what is called Western bourgeois and capitalist tyranny. The Western powers, on the other hand, are reluctant to move against a government ruling its own people tyrannically, or even against such a government extending its hegemony by armed invasion of neighbouring states. This reluctance stems only partly from a weakening of the earlier aggressive moralism; it stems chiefly from the traditional 'just war' condition of proportion. The harm likely to follow, in an era of nuclear weapons — and even of highly destructive 'conventional' weapons — from armed intervention to repel the invader would be so disproportionate to any foreseeable good that the invaded and oppresssed must be left to their fate. Sustained protest, and the mildest of cultural and economic sanctions, are the most that such an aggressor has now to fear — unless he threatens one of the Western powers directly. In other words, one of the legitimate grounds for war in the old 'just war' tradition, to right a wrong, is now eroded, partly by a blurring of distinctions between political right and wrong — athletes from seventy-odd nations are prepared to compete in the Moscow Olympic Games while their contemporaries in Kabul are shot in the streets or imprisoned — partly by the disproportionately destructive power of modern war.[6] The Christian is left with his paradox: he belives in a righeous God, whose righteousness must be vindicated; he must hope, therefore, and work, not despair; yet he cannot see how or when the vindication will come, and he knows that there are ways by which he ought *not* to try to hasten it.

One strand in the 'just war' tradition, the consideration of proportion, has been mentioned. With it goes another, the consideration of discrimination. In earlier centuries it was possible to distinguish with workable clarity between combatant and non-combatant; and so to regard an attack upon and disablement of the combatant as a legitimate act of war, but to forbid direct attack on non-combatants, not contributing to the enemy's effort in war. Today such clear distinctions cannot be drawn, partly because of the wider range of persons mobilised in war, partly because of the destructive power of modern weapons. In particular, nuclear warheads used 'strategically' — that is for the obliteration of the enemy's cities and total war potential — cannot be used discriminately, and therefore may not be used at all without grave moral offence. (The discriminating use of 'theatre' weapons, like the enhanced radiation warhead, is a matter for further discussion, involving technical issues of its own.) The present theory of nuclear deterrence — by which, it is maintained, nuclear powers are restrained from launching a nuclear attack by the fear of an immediate nuclear counter-attack in return — rests not only upon the possession of such weapons of undiscriminating mass destruction but also upon a preparedness and a will to use them in response to such an attack. If the preparedness or the will to use them should fall into doubt on either side, the deterrent power is lost. It is frequently asserted, as a point of Christian moral doctrine, that a conditional intention to use immoral means is immoral as would be the use itself. And the theory of nuclear deterrence is discredited on this account.

But is this so? The intention in possessing a nuclear 'deterrent' force is *not* to use it, but to restrain a potential enemy from a first, provocative use. The *intention* is, by maintaining a credible threat, to prevent any occasion for its use — to deter the other side from the first, immoral, act, the nuclear strike. If, knowing the consequences, he commits that act, the responsibility for the consequence is primarily his, however much the respondent also is to blame: he brings undiscriminating destruction upon his own head, and on all his people. While he is deterred from evil — a nuclear attack — by the threat of nuclear retaliation, the maintenance of that threat could be morally justified. To deny justification for carrying out that threat would be to rob the deterrent of its force: it is essential to its effectiveness as a deterrent that the other side should not discount the possibility that it would be used against him in return. This is not a bland assertion of moral innocence declaring itself against stark iniquity. As the deterrence is mutual, so also is the moral responsibility: whatever the degree of moral obloquy attaching to the first nuclear strike, it will be incurred by the power which strikes first. Neither is it to foreclose the military question, whether the nuclear deterrent should be kept or disposed of on expedient grounds. Neither is it to assert that to hold the deterrent, or to use it, is a 'Christian' act. The problem is one of those tragic necessities which, as has been said above, cannot be categorised at all in Christian terms. There is no *Christian* solution to it. There is only a choice among evils; and there is the Everlasting Mercy for those who, in good faith, are driven to choose.

If the argument sketched here has any validity it would point to certain limited conclusions.

(i) Christian groups of practitioners and commentators, like those in CCADD, should not be diverted from their established method of work by ecclesiastical or pious complaints that their reports are not 'Christian' enough; they stand in the tradition of Christian men who have wrestled with the problems of politics in the face of convulsive changes in the world — the vicissitudes of life in and after the Roman Empire; the testing of the 'just war' doctrines in the wars with Saracens, Turks, infidels and heretics; the emergence of national states; the conquest of the New World; the development of military technology; the establishment of the Nazi and Stalinist tyrannies.

(ii) Christian idealism, founded in faith, gives us goals; while Christian realism, founded also in faith as well as in experience, dictates realism as to means, especially the duty to deploy and control effective power.

(iii) If appeal be made to the Bible, the Bible must be used with the utmost exegetical integrity — a condition which excludes the uncritical extrapolation of words and acts from the theological context of the mission of Jesus and the experience of Him in the primitive Apostolic community to the political context of our own day — just as it would exclude a like extrapolation from the calls to exterminate the enemies of YAHWEH in a holy war which occur in the Old Testament.

(iv) The Christian Gospel will be most effective in this political context in the character which it imprints upon Christian men carrying responsibility in the relevant exercise of judgement and use of power — notably in the gifts of penitent self-knowledge, humility, patience, courage; and, perhaps above all, prudence. 'A right judgement in all things' is the substance of an ancient Christian prayer; and, indeed, of a Dominical promise.

Notes

1. Galatians 3:28: 'There can be neither Jew nor Greek, there can be neither bond nor free, there can be no male or female for ye are all one in Christ Jesus.'
2. Colossians 3:11: 'Where there cannot be Greek and Jew, circumcision and uncircumcision, barbarian, Scythian, bondman, freeman: but Christ is all, and in all.'
3. F.H. Russell, *The Just War in the Middle Ages* (Cambridge University Press,

Cambridge, 1975).

4. Walter Ullmann, 'Public Welfare and Social Legislation in the Early Medieval Councils' in G.J. Cuming and D. Baker (eds.), *Councils and Assemblies* (Cambridge University Press, Cambridge, 1971).

5. Haddon Willmer, 'Forgiveness and Politics', *Crucible* (July-September 1979), pp. 100-5; Sydney Bailey, 'The Christian Vocation of Reconciliation', *Crucible* (July-September 1979), pp. 126-30. Cf. G.R. Dunstan, 'Forgiveness: Christian Virtue and Natural Necessity', *Theology*, LXX (1966), pp. 116-20.

6. The calculation of proportion, both in the decision whether to go to war and in the conduct of war, is a task in which expertise and responsibility attach to the relevant knowledge. For the place of such calculation in ethics see G.R. Dunstan, 'The Ethics of Risk' in Nicholas A. Sims (ed.), *Explorations in Ethics in International Relations* (Croom Helm, London, 1981).

3 A CHRISTIAN UNILATERALISM FROM A CHRISTIAN BACKGROUND

Bruce Kent

A Christian Unilateralism

Before looking at the issues of unilateralism, I would first like to do a little scene-setting from what is at least one Christian point of view. Without the slightest wish to de-church anyone who happens to think otherwise, it seems to me that if Christianity involves general attitudes to war, violence, militarism, evil, hatred and to the nation-state, then it ought to be able to help us in the 1980s in our judgements on the arms race and our role in bringing it to an end. It will not be necessary therefore to justify the 'Christian' of the title. Not for one moment is the use of this qualification meant to imply a moral judgement on anyone else or to suggest that all Christians have to come to exactly the same conclusions on every 'defence' issue. Remarkable indeed if this were to be so.

Christian faith ought, nevertheless, to influence Christian judgement on 'defence' issues as much as, and indeed sometimes more than, considerations of strategy, economics, psychology and military technology, important as these things are in any discussion about disarmament.

It is often, and correctly, said that Christ did not found a political party, that he offered more than a political vision and that the Gospels were never intended to be a set of rules and regulations designed to guide twentieth-century human beings through all the moral difficulties of the day. Of course not. To be told that we are all members of the Body of Christ does not at once answer all of today's problems about immigration, racial harmony and unemployment. But unity in Christ does give us a particular starting-point.

We are often told that Christ was not a revolutionary or a political leader, all of which is perfectly true, if we define politics in a limited organisational way and if we think of revolutions only in terms of armed uprisings. Nevertheless, it seems just as clear to me that Christ who gave us the Word also gave us a vision of the Way. To be a Christian is not just a matter of signing up for a set of doctrinal propositions, but more importantly it is to become part of a living Body with its own norms and values in which the hidden Kingdom grows, inevitably, in conflict with the very different values of the world. Indeed, if such conflict does not exist then it is likely that we have compromised Christianity so that its challenge has been lost in conformity and introspection.

The Body of Christ, which of course includes the institutional churches but is a much larger, if less measurable, community than any of them, has its own standards, and they are not the standards which win the approbation of the world. Power, wealth, success, dignity, physical survival at any cost, extreme nationalism, violence, whether of the state, of other structures or of the individual are not the norms of Christianity. Christianity, and any religion actively concerned with the way people live as a loving community, has to have a political context. It exists incarnate but out of step with accepted patterns and is at once a challenge, a threat and a shining hope.

The profoundly political implications of the 'Our Father', if seriously meant, are quite clear. A common Father and a shared human family must involve critical attitudes towards many of the divisions, geographical, social, racial and economic in world society which we have too often come to accept as normal.

In the introduction to their document *Justice in the World* which the Roman Synod of Bishops produced in 1971, a remarkable passage occurs: 'The Church, the people of God, is called by God to be present among all men, preaching good news to the poor, freedom to the oppressed, joy to the brokenhearted. This gospel has the power to make men free not just from sin but from what sin has done to our society.' In a further section on education comes the striking self-criticism that we Christians have been so often preoccupied with the established order of things that all we have managed to produce is 'a carbon-copy image of that order . . . a far cry from the new man of the gospel'. None of this gets us out of facing the hard issues and current moral and political dilemmas of the world of today but it does give us a framework, which is not everyone's framework, as we struggle with the problems which face us. It would be hard to follow the Gospel and not to judge that ultimately history is the Lord's and that we are no more than short-term stewards, partners with the Lord in his ongoing work of creation. The book of Genesis is not a freeholder's charter. It is the start of the process of God's love and care which puts no person or group of persons beyond the pale.

How does all this apply to the problem of 'defence' in general and of unilateralism in particular? If there is no application, then we have indeed created a sacristy religion. But there must be. Christianity commits us, for instance, to the reality of internationalism as a priority over nationalism: to the priority of non-violence over violence: to the idea that it is better to suffer injury than to inflict it: to a belief that social justice can be the only basis for true peace: to an acknowledgment that we are stewards only of all we have and that what we have we hold on trust: to an obligation not to ignore the reality of evil in myself or in others, but to have also a heart of compassion, of reconciliation, of forgiveness and of understanding of the fears and motives of others, especially 'the enemy', in the belief that love and sacrifice are the only effective paths to real community.

Now all this may sound like the warm-up to a Sunday sermon strong on rhetoric but weak on application. I hope that such is not really the case. Every one of the social attitudes which I think follow from the Gospels, some of which I have tried to list above, should at least be brought to bear on contemporary 'defence' problems. Perhaps Christians will still not come to agree on matters of detail, but I doubt if they can really be so far apart on issues of principle.

Take two practical problems — our attitudes to nationalism and our attitudes to violence. The Gospel says something to both. Extreme nationalism, evident in liberal democracies and in totalitarianisms of the left and the right alike, is offered by the world as a form of unity of more importance than the unity of the Body of Christ. It is curious that there has been so little theological reflection on this contemporary phenomenon in the name of which so much is demanded. 'Standing as I do in view of God and Eternity I realise that Patriotism is not enough. I must have no hatred or bitterness towards anyone.' So said Edith Cavell before her execution in October 1915. This is the thinking of the Gospels. The tower of Babel was to be replaced by the Unity of Pentecost. Christ gave a new dimension to the idea of neighbour. The distrusted foreigner, the Samaritan, was to take on that role. St Paul, while not denying local loyalties which are obvious and valuable, explained that in Christ there was to be neither Jew nor Greek, slave nor freeman, male nor female (Galatians 3:28).

This was not an easy lesson. We know with what difficulty it was grasped by the Apostles that the good news, on equal terms, was also for the Gentiles. Nationalism today is not the same thing as the tribalism and religious insularity of those days. It has become even more extreme and its demands even greater, yet Christians seem little aware that a world system that involves over 150 nation-states, each separate and sovereign, presents problems which are not only political, strategic, economic and military, but theological as well.

So also is it in our thinking about violence. As an absolute minimum, a Christianity true to the Gospels and to the faith and practice of the Christian community must have the most restrictive attitude to violence, be it personal or collective, in the pursuit of justice. If ever it is still permissible for the Christian non-pacifist, war is a last resort hedged about with restrictions and conditions which it may be very hard now to meet. Said Pope John XXIII in *Pacem in Terris* (1963), 'in this age which boasts of its atomic power, it no longer makes sense to maintain that war is a fit instrument with which to repair the violation of justice'.[1]

For Christian pacifists there is an even more absolute position: violence is not a possible Christian option. St Cyprian in the second century was only one of the early Fathers who had things to say on this issue. 'The world,' he said, 'is wet with mutual bloodshed and homicide is a crime when individuals commit it, but it is called a virtue when many commit it. Not the reason of innocence but the magnitude of savagery assures immunity for crimes.'[2] Most Christians are not pacifists, though more and more are becoming such, inspired by the active non-violence of so many contemporary Christian heroes.

But the Christian non-pacifist has also to stand apart from the world in which he lives. He cannot make national survival an absolute good and he has, if tradition is to be taken seriously, to admit that total war can never be a Christian option. More than that, Christians of all traditions have to take seriously the possibility that the way of the Cross is actually a practical political option and that faith as well as prudence may suggest that there are times when 'it is better to suffer injustice than to defend ourselves'.[3] Loving our enemies may mean surrendering our lives or our liberty rather than taking theirs. It would not be hard to show from history and from the contemporary world situation that Christianity often grows and becomes more authentic under persecution.

Unhappily, the vision of a Christian alternative set of values on any of the major social issues is not one that has actually widely operated in practice; Caesar has taken much more than is his rightful portion, and in practice Christianity, wherever planted, has become a religion of conformity with the military and other policies of the nation-state. With honourable individual and collective exceptions we have not become famous for our struggles for social justice, for practical works of international reconciliation or for our interest in the still embryonic forms of international life. The contrary is perhaps true. Professor John

A Christian Unilateralism

Ferguson in his *War and Peace in the World's Religions* had a hard concluding sentence on Christianity:

> For the majority of the Christian laity at all times the claims of their nation have been paramount . . . The historic association of the Christian faith with nations of technological advancement has meant that Christian peoples, although their faith is one of the most pacifistic in its origins, have a record of military activity second to none.[4]

The non-application of the message of the Gospels is not only confined to problems of peace and war.

For many Christians the corporate social consequences of membership of the Body of Christ are not a principal perspective. The yeast in the dough is too easily understood in terms of individual conversion than of a corporate life with political and economic consequences.

Us and Them

The vigour, not to use a less charitable word, with which unilateralists are frequently attacked has often made me wonder if what is at issue is not particular independent initiatives designed to reduce weapon levels or to improve the international climate, but rather different concepts about world society itself. In particular I suspect that a polarity has been created in the world which ought to be unacceptable to Christians. That polarity is between 'us' in the West — democratic, free, non-expansionist, decent people — and 'them' in the East — malign, untrustworthy, aggressive and cruel.

This is, of course, a caricature, but not too much of one. It leads Christians here to accept whatever comes from 'our' side as credible and reliable, but from 'their' side to be part of a long-term plot. Once more we are fighting out, in hot wars and in cold ones, the battle of Good and Evil. It is a long time since Pope Urban II in 1095, in launching the first Crusade, sent off his knights to 'exterminate this vile race',[5] but the spirit of judgement and of polarity has not changed.

Before getting on to specific suggestions about independent unilateral initiatives which might in different ways do something to stem the present costly and dangerous arms race, perhaps we ought to look at the polarity issue a little more closely. It is an area to which Christians ought to bring some judgements from their own religious sources. We know that we are not angels and it ought to be a reasonable presumption that guilt for an arms race is not likely to be wholly on one side. It ought also to be an assumption that 'the other' may also have fears and perceptions which we ought to be trying, because of our mission as reconcilers, to understand. It might be reasonably assumed also, and all history provides the practical evidence, that our leaders too are not strangers to the love of power or unaware that nothing unites a people more than an outside enemy full of evil intentions. Real threats can, of course, exist but they can also be deliberately exaggerated out of all proportion. It would be as foolish to distrust everything from 'their' leaders as it would be to accept without examination everything that comes from 'ours'. This is not just a matter of practical political caution. There are theological issues involved in allowing ourselves to split the world into dark and light. I do not believe that it is a process which Christians can accept. It is not true to any belief in the ultimate goodness of humanity or the working of God's grace, and it renders us vulnerable to easy identification with one side in what is clearly a dual process of fear and counter-fear as mirror image meets mirror image and the arms race proceeds apace.

That we are so affected is hardly a matter of argument. The American report NSC 68, prepared in 1950, set out in words what has been in practice the operational doctrine of the West since the Second World War. 'The Soviet Union, unlike previous aspirants to hegemony is animated by a new fanatic faith antithetical to our own and seeks to impose its absolute authority over the rest of the world. Conflict has therefore become endemic.'[6] So it has been in an almost uninterrupted process since.

An American ex-diplomat who served in the Soviet Union at a critical time at the end of the Second World War now acknowledges the dangers of that process. In 1974 George Kennan wrote that the idea

> that the Soviets operate on totally different principles is an example of . . . a peculiarly American tendency to what I would call dehumanisation of any major national opponent: the tendency that is to form a species of devil image of that opponent, to deprive him in our imaginations of all human attributes and to see him as totally evil and devoted to nothing but our destruction.[7]

This kind of thinking, well described by Kennan, is not only bad politics, it is also bad Christianity. It reserves God for 'our' side, and in extreme forms it actually leads people to such a crusading fervour that they imagine that somehow there is a justified relationship between quite barbaric weapons of mass destruction and the defence of the message of Christ.

The easy division of the world into the good, defensive, freedom-loving West and the malign, expansionist, militaristic and tyrannical East just does not hold up to any serious examination of the history of the arms race, of post-Second World War military activity by the superpowers or of the character of the states which make up the *de facto* Western military network. The different judgements passed on superpower action in Afghanistan and in Vietnam, in Chile and in Czechoslovakia, should be obvious to Christians who ought to be consciously trying to operate on one yardstick. Something must be wrong with the theory of the good West when we find ourselves supporting the brutal military tyranny now existing in South Korea and defending the right of the Kampuchean Pol Pot regime to a seat at the United Nations so as not to offend our new Communist friends in China.

Disarmament initiatives are also seen in the same bifocal way. President Carter's remarkable, though coming in the middle of the SALT II process rather startling, proposals in 1977 for deep mutual cuts in nuclear arsenals, foolishly rejected by the Soviets, became incontrovertible evidence of Western good intentions. Mr Brezhnev's proposals of February 1981 for immediate direct discussions with the United States on the limitation of arms, for limiting the deployment of new submarine systems and for a freeze on new medium-range missiles were, by contrast, either ignored or treated as yet further evidence of Soviet guile.

It is not surprising that an endless propaganda campaign in the West to prove Warsaw Pact military superiority in nearly everything succeeds so easily. It is difficult to remember that as recently as autumn 1980 the Pentagon was saying that US military strength remains 'second to none' and that charges to the contrary are based on 'misleading statistics' and 'rhetorical nonsense'.[8] It is hard to know which information process is the less satisfactory — the American habit of offering a great deal of information and wildly different interpretations of it, or the Soviet style of oppressive military secrecy.

Our own British authorities have also engaged at times in the process of manipulating public opinion and have not been unwilling to stoop to low levels when it has been thought appropriate to smear the peace and disarmament movements as agents of the Soviet Union. The allegation has even been made recently by a senior parliamentarian, and echoed in the House of Lords, that 'the peace movement' is 'organized and financed by Russia'.[9]

The April 1981 White Paper told the public that 'the combination of geography and totalitarian direction of resources gives the Soviet Union a massive preponderance in Europe'. Since this is manifestly untrue of nuclear weapons, the preponderance must relate to conventional forces. Upon what evidence this opinion is based, how one is meant to assess military balance or what non-military factors, for instance the political reliability of alliance partners, should be taken into account is not made clear: it certainly ignores the China factor, that great threat to the Soviet back door.

Christianity gives no one any special competence in such a work of assessment, but it ought to help the individual Christian to take a detached attitude even to military assessments coming from 'our' side. The more so in that sometimes information is quite deliberately suppressed as in the case of Chevaline – (the secret billion-pound British programme for 'improving' Polaris missile warheads so as to enable them to penetrate the defensive systems of Moscow) – or *The War Game*, the nuclear war film still denied to the British public. Sometimes standards sink lower and the British government produces information which is not only bland but clearly misleading, as happened in the case of what might well be called the cruise-missile sales brochure issued by the Ministry of Defence to reassure the residents of Berkshire and Huntingdonshire, where it is suggested that the new missiles be based. In the brochure's estimate of theatre nuclear systems the West is shown to be at a major numerical disadvantage by simply making no mention of NATO submarine or sea-borne weapons or of French nuclear capacity. The brochure neglects to state that the cruise missiles proposed for this country are to be under exclusively American power to operate. Finally, it gives assurances to local residents, entirely belied by the assumptions of the subsequent 'Square Leg' civil defence exercise. In that exercise the cruise missile bases were assumed to be major targets.

None of which is meant to suggest that there are not also in the corridors of military power on 'our' side honourable people facing terrible dilemmas. Nor is it meant to suggest that opinion-forming is a British or a Western speciality. On the contrary, it is universal practice and the Christian ought to be critically aware of it. The links between the churches and the military establishments of the different countries, including those of the East, which make the churches especially vulnerable are subject-matter enough for a different paper. It is enough at this stage to stress once more that membership of the Body of Christ should not blind one to cruelty and violence anywhere, but should

A Christian Unilateralism 61

make Christians a little more detached than they often are from national perspectives.

Unilateralisms and Other 'isms'

At a recent wedding I was approached by one of the aunts of the bride who noted my CND badge. Unusual for weddings, the subject of nuclear weapons started up. She came quickly to the point. In total conviction she told me, as if she was talking of absolute chemical opposites; 'Ah, but you are a unilateralist — I am a multilateralist.'

One of the most effective campaigns which has been waged against the disarmament movements in recent months has been the one which has produced this apparent gulf between multilateral negotiations and unilateral actions. The first is respectable disarmament procedure. The second, its opposite, is imprudent, ill-informed and recommended by well-meaning but foolish people often manipulated by those not quite so well meaning.

It is well, therefore, to start this final section with the assertion of a contrary view. Bilateralism, trilateralism, multilateralism and any other form of international disarmament negotiation are in no way in conflict with unilateralism. I know of no unilateralist who believes that, as a matter of practical politics, it should be recommended that all military forms of power be abandoned forthwith. Such a proposition would be doomed before it started. Those of us who are pacifists know this perfectly well. We work in a culture and a political system which deeply believes in the value of what is thought to be military protection. Unilateralism has been made, by its detractors, to sound like the imposition of instant pacifism and, were that a proper description, then it would deserve most of the criticism that it gets.

But it is, of course, not so. There is first of all no radical division between the different processes of disarmament activity. The Final Document of the United Nations Disarmament Session of 1978[10] refers no less than three times (paras. 27, 41, 114) to unilateralism in the context of other normal procedures. The unilateralist therefore has no quarrel with those who urge multilateral, bilateral or regional negotiations. Many unilateralists take part in them and see them as entirely complementary to their unilateral concerns. Hesitations only arise when multilateral agreements are portrayed as being more significant than they actually are and so have the counter-productive effect of reducing interest in genuine disarmament. It is no place to make that case here but there are, for instance, those who would judge the Partial Test Ban Treaty and the Non-Proliferation Treaty in that light. Nevertheless, though the former has not prevented nuclear weapon testing, it must have saved many from death from the results of nuclear fallout. The latter has not prevented nuclear weapon proliferation and is not likely to do so, if the nuclear powers continue to neglect their own obligations under it. It is, nevertheless, at least international official recognition of a major threat. It is arguable that had the (bilateral) SALT II agreement come into force it might itself have had more of a cosmetic than a real effect.

None of this is to diminish the value of international negotiations. There is no reason, however, why at the same time each state (and indeed each church, university, trade union, newspaper, aid agency, political party, cultural group, etc. *ad infinitum*) should not also be seeking to take such steps as are within its power, without getting the permission or agreement of any other body in advance, which may reduce tension, increase confidence and even lower the level of armaments. That kind of direct initiative in a conflict situation is perfectly normal in personal quarrels and in all conflict situations below the nation-state level. Above that level, for reasons which escape me, it is thought to be a hazardous and imprudent procedure, even in a world in which weapons systems now far exceed possible targets.

Which unilateral actions may be possible and appropriate must depend entirely on the context. They will differ according to the positions of different states, according to the perceptions and influence of public opinion in each state and, of course, according to the alliance systems and weapons level of each state. It seems to me to be entirely reasonable for someone to suggest that a particular unilateral proposal has been badly carried out and has not therefore had the confidence-building effect that might have been anticipated. This might perhaps be said of the modest Canadian commitments not to carry or stockpile nuclear weapons made after the 1978 Special Session. It is also entirely reasonable to hold that a particular unilateral initiative may be, for a variety of reasons, quite inappropriate and may perhaps increase rather than reduce risk.

Each unilateral proposal must be looked at on its own merits by those concerned with the reversal of the arms race, but it makes no sense to suggest that no unilateral actions can ever be contemplated. Some norms for judging the appropriateness of unilateral activity have been prepared by Charles Osgood and were referred to by Sydney D. Bailey on page 7 of his statement commending the British Council of

A Christian Unilateralism

Churches resolution (19 November 1979) on *The Future of Britain's Nuclear Deterrent*. Others might be added. That there may be risks in unilateral processes, no one would deny, but any assessment of risk must also take into account the dangers of the present situation. The point was well made by the 17 Bishop members of Pax Christi, USA in March 1981, who declared that 'Granted, unilateral initiatives involve serious risks, but in our view these risks are called for in the light of the far greater risks of an arms race which the Vatican (in 1976) has called a machine gone mad.'[11] Pope John Paul II, in his World Peace Day Message for 1979 acknowledged this also by implication, 'Make gestures of peace,' he said, 'even audacious ones, to break free from vicious circles and from the dead weight of passions inherited from history. Then patiently weave the political, economic and cultural fabric of peace.'[12]

But what gestures in the context of the United Kingdom? I do not think that the Pope had only governments in mind. Too often the churches see themselves as sitting in judgement on the morality or otherwise of governments, as if that was their only role in peace-making. On the contrary, it is a rather more Christian perspective that only as instruments of peace can we bring peace. A wealth, therefore, of possible unilateral church activity lies behind the statement of the Dutch Roman Catholic Bishops of 1968, which concluded: 'Looking for peace means giving peace a real place, not only as a pious wish in our hearts and on our lips, but in our thoughts, in our interests, in our educational work, in our political convictions, in our faith, in our prayer and in our budget.'[13]

Similarly, at British government level the range of possible unilateral activity is very wide indeed. That there has been an increase in public awareness about the risks of the arms race has been due very largely to the work of voluntary non-governmental organisations. Yet in the 1978 United Nations Final Document (paras. 100-6) it was agreed that governments should take their responsibility for this vital work of information. Public education could, for instance, take the form of school and university programmes, of support for the annual UN Disarmament Week, of increased contact between government bodies and the peace movements, of government support for effective media initiatives. As it is, the ordinary citizen receives far more information funded from government sources about the risks of smoking than he or she does about the arms race — a much greater health hazard.

Unilateral actions related to education are not the only ones that can be easily imagined. Government support for, and interest in, redeployment schemes proposed by unions or union groups working in military industry would be a most constructive gesture of wider vision. So also would be a much more rigorous approach to the national arms export industry. As a minimum, in a democratic society, we should be able to know in advance what deals are being contemplated and why, who is to be invited to arms 'fairs' and where weapons systems eventually in fact end up. For the Christian non-pacifist, as for the pacifist, this is not an industry just like any other. Selling weapons is not exclusively an economic issue.

Further, it would be most constructive if the government took seriously the investigation of alternative defence systems. There is a range of literature both on alternative military and non-military systems, yet the idea that 'having more' so as to be able to 'negotiate from strength' still seems to be the level of much official thinking. The ineffectiveness of our own army in Northern Ireland in the spring of 1974 in the face of a mobilised and obstructive population ought to have lessons for us. The United Kingdom had made its political will clear, but the military ability to enforce a power-sharing policy was simply lacking. The Vietnamese and the Afghans now are both examples of the inadequacy of the view that greater military strength is enough to guarantee political power. This is no place to introduce an essay on alternative defence, military or non-military. But it is not too much to ask that the subject should at least have some serious official attention. Mass civilian non-cooperation, economic pressure, planned guerrilla warfare, and activity aimed at destroying the morale of an aggressor's forces are all at least elements in 'alternative' thinking. Sir Stephen King-Hall's *Defence in the Nuclear Age* was a pioneer contribution to the alternative defence discussion on which there is now a large literature.[14]

These points I only make to show that unilateralism need not only relate to specific weapons systems — confidence-building has a much wider context. Nevertheless, in the context of weapons, most if not all British unilateralists would be united in opposition to both 'independent' British nuclear weapons and to the further escalation of American military might in these islands represented by cruise missiles. At the same time, most would be wholly in favour of agreeing a nuclear 'no first-use' pact with the Soviets. It is a national scandal that our own military policies will almost certainly lead us to use nuclear weapons first in an East-West conflict, while at the same time many of the most serious military thinkers of our time, like the late Lord Louis Mountbatten, reject the idea that nuclear war can be contained at some

A Christian Unilateralism

limited level below that of a general holocaust. Unilateralists might well be more divided about the political wisdom of immediate withdrawal from NATO (not that this means sheltering under anyone's umbrella, since umbrellas are exactly what the superpower arsenals are *not*). They would certainly urge on other countries reciprocal actions, but not actions without which our unilateral steps would not proceed. This is particularly the perspective of the European Nuclear Disarmament campaign, which was launched in 1980 as the result of an initiative of the historian E.P. Thompson. This campaign is aimed at establishing a nuclear-free zone in Europe, in the phrase of its first declaration 'from Poland to Portugal'. It has already launched a number of European initiatives.[15]

The arguments for abandoning a British independent nuclear role are various on the strategic level. They are incredible as a serious threat to the Soviet Union and do not, in fact, present a real second centre of decision. For a threat to be credible it must be able to carry conviction that the threat will in given circumstances be carried out. Unless we are committed to a policy of independent national suicide on this small island, Polaris and Trident remain symbols rather than threats to the supposed Soviet aggressor. They have not earned us a place at the top negotiating tables. Were it otherwise, we would surely be partners and not spectators in the SALT process. They are a standing argument for proliferation and, especially in the Trident context, vastly expensive and represent a policy funded on a willingness to commit wicked acts beyond imagination. This is perhaps, late in the chapter, *the* argument for nuclear unilateralism of major Christian appeal. For too long we have failed to look at the actual situation because we have either been distracted by hypothetical scenarios in which nuclear weapons might be used in a discriminate way or by arguments related to conditional intentions. There are no conditional intentions in this matter. The willingness to use nuclear weapons is a present reality on a daily basis for those in the armed forces who have responsibility for them. However strong may be the conviction on their part that this is the way to peace, the reality is that it only requires an order to be given for a trained and tested obedience to operate. Nor is there any way in which our present strategic nuclear weapons can be used discriminately even if limited nuclear war were actually possible and the genetic damage to future generations could be ignored. They are not sufficiently accurate and indeed, by a strange incongruity, the more accurate they become the more they will be seen as the weapons of pre-emptive first-strike and the more unstable the situation becomes.

The opposition of unilateralists to cruise missiles runs on rather different lines. Their accuracy, yet their comparative slowness, creates a problem as to how they should be defined and as to what they are for. Perceived by the Soviets as a strategic threat, they present us also with considerable additional dangers. Professor Michael Howard, arguing for a more credible civil defence programme, wrote in January 1980, that 'the presence of Cruise missiles on British soil makes it highly possible that this country would be the target for a series of pre-emptive strikes by Soviet missiles'.[16] He is quite right: they are exactly the kind of target that the Soviets claim for their SS.20s. They cannot add to deterrence, they take us closer to war-fighting, they are a further addition to the world's nuclear arsenals, and the unilateralist would certainly reject them, and with them, all nuclear weapon developments which lower the threshold between nuclear and conventional warfare and create illusions about fighting and winning limited nuclear wars.

However, I am not so much concerned with these two particular unilateral issues as with establishing the validity of the unilateral process in general. It is a process which, of course, applies in different ways to different countries. Immediate non-nuclear status makes sense for the United Kingdom. It does not for the United States or for the Soviet Union, not because there is some difference in the morality involved, but because, granted the perceptions that exist, it is an unrealistic political objective. What is possible for both superpowers is immediate deep cuts unilaterally in their nuclear arsenals. Both, in terms of deterrence, have an irrational overkill capacity.[17] Both could make substantial reductions and start a unilateral process which in their case has to be, for practical reasons, different from our own.

It is the validity of unilateralism in general which is being defended. If that is agreed, then we face only discussion about the appropriateness of particular steps open to churches as well as to governments. Not that there may be too much time for leisurely discussion. 'Mankind is confronted with a choice. We must halt the arms race and proceed to disarmament or face annihilation.' So said the 1978 UN Special Session Report (para. 18). In this task I would not claim that Christians have some special wisdom which guides their every step, but they do have a spiritual framework which makes reconciliation a first responsibility and internationalism a major perspective. If they differ in their views about the appropriateness of particular steps on the disarmament road, nevertheless, as members of the transnational Body of Christ, they are committed to the priority of the work for peace, a work which will often challenge the assumptions of nation-states.

Notes

1. *Pacem in Terris* (1963), 1st edn, S. 264 (Catholic Truth Society, London, 1967), para. 127.
2. Quoted in James O'Gara, *The Church and War* (US National Council of Catholic Men, Washington, DC, 1967).
3. Quoted in the 1976 Vatican submission to the United Nations, *The Holy See and Disarmament* (Pax Christi International, Utrecht, DP/1976/8e).
4. John Ferguson, *War and Peace in the World's Religions* (Sheldon Press, London, 1977), p. 122.
5. Quoted in James O'Gara, *Church and War*.
6. Quoted in R.J. Barnet, *'New Yorker'*, 27 April 1981.
7. Ibid.
8. Washington Centre for Defence Information, *Military Information*, 3 October 1980.
9. See Bruce Kent in *Sanity* (October 1981).
10. UN Document A/Res/S.10.2, 30 June 1978.
11. *News Notes*, vol. 6, no. 4 (July 1981) (Maryknoll Justice and Peace Office), para. 20.
12. Pope John Paul's *World Peace Day Message for 1979* (text from Pax Christi, London).
13. Bishops of the Netherlands, Peace Message 1969, 'On Banning War' (Pax Christi International, Utrecht, DP/1969/5).
14. See, for example: Sir Stephen King-Hall, *Defence in the Nuclear Age* (Victor Gollancz, London, 1958); Dr Gene Sharp, *Politics of Non-Violent Action* (Porter Sargent, Boston, 1973); Adam Roberts (ed.), *The Strategy of Civilian Defence: Non-violent Resistance to Aggression* (Faber and Faber, London, 1967); R. Boserup and A. Mack, *War without Weapons* (Francis Pinter, London, 1974).
15. See END *Bulletins* (from END, 6 Endsleigh Street, London WC1).
16. Letter in *The Times*, 30 January 1980.
17. George Kennan, 'Einstein Peace Prize Address', *Disarmament Times* (June 1981).

4 IN DEFENCE OF DETERRENCE

Arthur Hockaday

CHURCH BANS NEW HORROR WEAPON! Had such a headline been fashionable in 1139 it might have greeted the prohibition by the Second Lateran Council of the use of crossbows against Christians (while presumably permitting their use against infidels and heretics). I cite this simply as a reminder that the churches have long concerned themselves with the questions that run through this collection of essays — questions with which we must all be concerned as Christians, as citizens and as human beings. But the surrounding climate of this concern has changed. For in the present century there have occurred two of those events after which nothing looks quite the same again. One was the outbreak of the First World War at the beginning of August 1914. The second was the dropping of a nuclear weapon on Hiroshima on 6 August 1945.

These two events have stimulated a secular attitude to war quite different from that which prevailed up to 1914. This has been particularly noticeable in the 'developed world' of nations which participated in one or both of the World Wars, or whose political identities were sufficiently formed to acquire a 'political memory' of them. It has probably helped to foster, and has in turn been strongly influenced by, a growing social awareness of the kinship and interdependence of mankind and a gradual reassessment of the purposes of political action.

Before 1914 wars had seldom involved whole nations. Their direct effect had been confined to the military classes, those who attached themselves temporarily to those classes and those whose land was ravaged or their possessions commandeered. Military glory had a certain glamour; war was regarded as a reputable pursuit for gentlemen; and serious attention was paid to sententious declarations that war could bring out the finest qualities of a nation's manhood.[1] Resort to war in 1914 appealed variously to desires for territorial and economic expansion, to frustrations at political or diplomatic impotence, to feelings of political obligation and to resentment at the perceived aggressive intentions of others. It was generally received with enthusiasm. We in Britain are familiar with photographs of straw-hatted young men cheering in the City of London. There exists a photograph of a similar crowd in the Odeonsplatz in Munich, in which the youthful Adolf Hitler has ingeniously been discerned.²

The carnage of the First World War changed all that. Some 13 million may have been killed in all. British casualties, with rather more than one million killed, were fewer than French, German or Russian in proportion to numbers mobilised;³ yet the 54,896 names on the Menin Gate at Ypres, representing simply those soldiers of the British Empire who fell in the Ypres salient and have no known graves, exceed the total number of American servicemen who died in Vietnam.

The Second World War was forced upon the world by a mad and evil tyrant, drawing with him the nation which he had inexplicably corrupted. Its outbreak provoked more sighs of weary resignation than cheering in the streets. The casualties were far greater than in the First World War — deaths probably exceeded 50 million, with some 20 million Soviet citizens alone perishing in battle or through deprivation. Yet they did not have the same impact upon a popular consciousness inoculated by 1914-18. It was 1945 that brought a shock of kind rather than of degree.

The Allies had devoted a massive concentration of resources to developing a nuclear weapon before the Germans could do so. By the time the new weapon was complete the German war was over; but it was now seen as a means of bringing the Japanese war to a sudden end without the need for an all-out onslaught on Japan or its accompanying destruction and casualties. How materially it shortened the war, and whether to do so it was necessary to destroy two of Japan's larger cities, are questions which it is still legitimate to debate. What is beyond dispute is that the demonstration of the effect of two relatively primitive nuclear weapons introduced a new dimension of horror to the world's perception of war.

The attrition and devastation of the two World Wars, the prospect of far greater devastation inherent in the multiplication and technical development of nuclear weapons, and in many countries the more direct accountability of rulers to their peoples, have created a climate in which, at least in the developed world, even the most absolute ruler is unlikely to have recourse to war as light-heartedly as in the past — Émile Ollivier's acceptance *d'un coeur léger* on 15 July 1870⁴ of the responsibility for initiating the events which led within seven weeks to the battlefield of Sedan was but a particularly egregious expression of a widespread attitude. No one who is conscious of the effects of conventional attrition or nuclear devastation will wish to risk incurring them. Hence he will be unlikely to have recourse to military force unless he can be confident not merely that he can gain a victory, but that the cost of doing so will be acceptable to him. Both parties to a dispute may be expected to share this basic approach. If so, military force is unlikely to be brought to bear unless the balance of strength between the two parties is so one-sided that the weaker cannot offer significant resistance, or unless the stronger mistakenly believes this to be so. The possibility that the stronger party may miscalculate either the nature of the balance or the resolution of his opponent means, however, that armed conflict cannot be ruled out even when he might have been expected to comprehend its implications fully. A recent example may be the apparent misjudgement by the Soviet Union of the duration and intensity of the resistance which its troops would meet in Afghanistan.

This concept of balance and imbalance is relevant to the use of armed force not only as a means of military aggression but also as an instrument of political pressure. This will be particularly relevant where the weaker has let the balance deteriorate to the point where, perhaps not until it is too late, he perceives himself as having no choice but to conform with the wishes of the stronger. If, therefore, we do not wish to be the victims either of military aggression or of undesirable political pressure, we shall seek to be able to deter any prospective opponent from using his armed forces to these ends. We may recognise also that anyone else will similarly wish to deter us.

Deterrence has thus come to be a principal function of armed forces and of military policy. It consists essentially in being able to make clear to any prospective aggressor, military or political, that we have a valid option, or a range of valid options, alternative to surrender and that we are prepared to resort to them. So long as the purpose of the deterrer's armed forces is the prevention rather than the promotion of conflict, and so long as his objective is no more than to preserve the *status quo* in territorial and geo-political terms, he does not require a military capability sufficient to inflict a defeat in the traditional sense or to acquire, destroy or hold down large tracts of territory. He must, however, be capable of inflicting damage that a potential aggressor will regard as an excessive price for whatever he may hope to gain, and of making clear to a potential aggressor that no easy victory is available to him. For this he must have a genuine military capability valid in combat. To draw a sharp distinction between the utility of military forces for defence and their utility for deterrence is misleading, for the deterrent properties of a military capability inhere in its perceived ability to present a robust defence or to effect a given amount of destruction. And this capability must be backed by evidence sufficient to persuade a potential aggressor of the defender's willingness and

In Defence of Deterrence

intention to use it if necessary; for a prospective aggressor will not be deterred if he believes that the defender is bluffing and is not prepared to use his military capability in defence or retaliation.

Deterrence is sometimes likened to a game of chess, in which an opponent's moves can be foreseen, reactions to them planned, and various options for both the attacker and the defender thought through. This analogy occurs in the section entitled 'Nuclear Weapons and Preventing War' in the British government's Defence White Paper of April 1981, to which Bruce Kent has referred in Chapter 3. The present essay expresses a purely personal viewpoint and should not be taken as necessarily reflecting that of Her Majesty's Government; and so I have reproduced the section from the White Paper in full as an appendix to this chapter, setting out in the government's own terms its analysis of what deterrence is about in the nuclear age.

Yet deterrence is more than a game, for the penalties of losing through any one false move are so much greater. For each move the nature of the capability to be deployed, and the prospective circumstances of its deployment, must be such as to render it credible that the will to deploy it will be exercised. There is a link here with the criterion of proportion traditionally regarded as a necessary criterion for the just conduct of war. In deterrence a particular move is likely to succeed only if the prospective action on which it rests is, both in itself and in its likely consequences, sufficiently proportionate to the action against which it would be a retaliation for an aggressor to find it credible that it will be taken. This last point is important for consideration of the place of nuclear weapons in deterrence. The term 'the deterrent' is sometimes applied exclusively to nuclear forces in a misleading sense which implies that they alone have a deterrent function. This in turn sometimes colours general discussion of deterrence to a point at which our natural fear of nuclear weapons leaves us in terror of our own deterrence. The use of nuclear weapons is, however, plausible only if the user has no fear of retaliation (as was the case when the United States used nuclear weapons against Japan in 1945) or if he is prepared to accept the damage that may be inflicted upon him in retaliation. This damage is unlikely to be acceptable if it involves extensive loss of population or economic assets or damage to other vital interests on a scale which we may describe as 'strategic'. It is difficult, and probably unprofitable, to attempt to quantify this with precision. General André Beaufre described in 1965 a French study in which it was estimated that a loss of between 2 and 10 or 15 per cent of a country's resources would bracket the damage that could be risked for most stakes, and that for no stake would loss of more than 50 per cent be acceptable.[5] Secretary of Defense McNamara judged in 1968 that a capability to destroy 20 to 25 per cent of its population and 50 per cent of its industrial capacity would serve as an effective deterrent to the Soviet Union as representing a level of destruction intolerable to any twentieth-century industrial nation.[6] As indicated later in this essay, I would regard the lower French figures as the more realistic in psychological terms.

I agree with Barrie Paskins in his assessment, in Chapter 5, of the present balance of strategic nuclear deterrence as robust. Neither the Soviet Union nor the United States can have any confidence of being able in a first strike to deprive the other of a capability to retaliate on a massive scale. Both superpowers, the United States in particular, have invested significantly in ballistic-missile submarines. Although anti-submarine technology may be expected to advance, only a dramatic breakthrough on a large scale could give any confidence of being able to destroy even a substantial proportion of the adversary's missile-carrying submarines; and the likelihood of such a breakthrough is officially considered remote.[7] Much has been said and written about the potential vulnerability of American ICBMs (inter-continental ballistic missiles) in the 1980s. The International Institute for Strategic Studies regards this as much more a theoretical than an operational concern. To destroy the 1,064 US ICBM silos, the Soviet Union would need some 2,000 perfectly co-ordinated warheads – including a second (and perhaps even a third) wave to compensate for failures in flight or on detonation in the first wave – all of them spaced and timed to avoid mutual destruction by the phenomenon known as 'fratricide', which can cause the nuclear explosions of some warheads to affect other warheads before detonation. Fortunately there is no empirical evidence to give any potential aggressor any confidence that such co-ordination can be attained; the view of the Institute is that the command and control requirements 'border on the infeasible' without taking account of tactical means open to the defender (launch-on-warning or launch-under-attack) which would leave only empty silos for incoming warheads to destroy.[8]

Strategic nuclear weapons are a vital deterrent, and in my opinion the only sure deterrent, to the use of similar weapons by an opponent. Their validity as a deterrent to any other category of aggression requires more careful analysis, for the first use of such weapons would be liable to incur retaliation unless their possessor had a monopoly of them. And what are generally called 'long-range theatre nuclear forces', a categorisation stemming from the limitations of range which make them irrelevant to any inter-continental exchange between the Soviet Union and the United States, are certainly perceived as capable of inflicting 'strategic damage' upon countries vulnerable to them. In the East-West context these are the European NATO countries and those of the Warsaw Pact, including substantial portions of the Soviet Union; and the Soviet Union has from time to time sought to include 'forward-based systems' within the scope of the Strategic Arms Limitation Talks.

These long-range theatre nuclear systems should not be confused with those of shorter range whose use would be closely linked with the tactical conduct of a battle and which are often discussed as a means of restoring deterrence should initial deterrence have failed, war have broken out and conventional forces be unable to check the aggression. As part of a strategy of 'flexible response' or 'graduated deterrence' a limited use of nuclear weapons is seen as a possible means, not only (or even necessarily) of restoring the tactical situation but of bringing home to the attacker the defender's determination to resist and his willingness to 'escalate' the level of conflict if necessary. On the assumption that the attacker will be reluctant to face the possibility of 'escalation' with its accompanying dangers for both sides, this use of nuclear weapons *in esse* as a form of 'intra-war deterrence' is conceptually similar to their use *in posse* as a deterrent to nuclear attack. Like any other form of deterrence, however, it involves a gamble on the probability that the adversary to be deterred will be rational; and history suggests that rationality tends to decline, and attitudes to harden, once an initial miscalculation or irrationality has sparked off a conflict. Moreover, although 'battlefield' or 'short-range theatre' nuclear weapons are not designed to inflict damage of the kind that I have described as 'strategic', they are unlikely to be used tactically by either an attacker or a defender unless the user judges that the associated 'collateral damage' to populations, land or buildings will be acceptable to him in proportion to what he expects to gain by use or to lose by refraining from use. And in considering 'collateral damage' we should not be misled by what may appear to be small 'yields' in numbers of kilotons. We measure the 'yield' of a nuclear explosion in terms of the energy released; and our unit of measurement is the amount of TNT that would have to be exploded to produce the same release of energy. Such an apparently small 'yield' as 'half a kiloton' (and this is indeed small in comparison with the approximately 15 kilotons of the weapon dropped on Hiroshima) means that 500 tons of TNT would have to be exploded to produce the same release of energy. This was the total 'yield' of the legendary simultaneous explosion of 19 mines which preceded Field Marshal Plumer's attack at Messines on 7 June 1917, from which some of the craters can still be seen today.[9] A decision to initiate the use of nuclear weapons in this mode would have to weigh carefully the balance of risk of retaliation, 'escalation' and 'collateral damage' against the consequences of acceptance of conventional defeat. It should therefore be a primary objective of strategy to ensure that such a decision does not have to be taken, or at least can be postponed as long as possible.

The role of nuclear weapons in deterrence must always be linked with abstruse calculations of loss or gain and with conjectures as to how an opponent will respond. There is scope for misjudgement of an opponent's willingness to bring his own nuclear weapons into play in a particular set of circumstances. Similar considerations apply to a much lesser extent to the role of conventional forces in deterrence. If conventional forces are physically deployed on the ground, on the sea or in the air, there is less ambiguity about their commitment; they must be physically engaged and brushed aside if territory is to be gained or hegemony imposed. A year or two ago I was walking through St James's Park and noticed a party of schoolgirls eating their picnic lunch on a visit to London. They were being harassed by a flock of birds who fancied their sandwiches. Suddenly one of the girls — and I quote her actual words — in her exasperation cried out: 'In polite language, bugger off!' I realised at once that in those five words this girl had encapsulated the essence of deterrence that I had been seeking for years. To be able politely but confidently to tell anyone who is harassing us exactly where he can go is as important in deterring the use of military force as an instrument of political pressure or blackmail as in deterring its use as an instrument of military aggression.

But for both confidence and deterrence a credible capability is essential. This is illustrated by the story of the Scottish rugby player Erle Mitchell who, after being lectured at length by the mighty Colin Meads of New Zealand on his provenance, his parentage and his future prospects, replied in precisely the terms recommended by my schoolgirl but 'didna say it in a verra loud voice'. This attempt at deterrence clearly failed, for later in the match the referee found it necessary to send Meads off the field for dangerous play.[10] In military terms the presence of conventional forces on the ground, on the sea and in the air can be the least ambiguous way of maintaining confidence and deterrence. To postulate conventional forces as in this sense a primary deterrent is in no way to call in question the validity of nuclear forces as an ultimate deterrent underpinning them; but the more robust the capability for conventional response, the greater the prospect of postponing or averting the painful nuclear decision.

A situation in which the prime utility of nuclear forces, and some might argue their sole utility, is to deter the use of similar weaponry by others is prima facie absurd. George Kennan has recently suggested that, since (in his phrase) the nuclear bomb can be employed to no rational purpose, it is the most useless weapon ever invented.[11] If so, why not burn the lot? But we have to recognise two basic facts. First, the relevant technology cannot be disinvented. Second, the only use of these weapons in history has been by a power which at the time was alone in possessing them. Unless, therefore, the most stringent and foolproof machinery could be devised for inspection and verification, their ostensible abolition would place an unacceptable premium on cheating and would tempt at least each of the two nuclear superpowers to cheat as an insurance against cheating by the other. It is not necessary to dispute that the present nuclear weapon inventories of both the Soviet Union and the United States are unnecessarily large in order to argue that there could be greater stability in a balance, not necessarily symmetrical, at a modest level than in ostensible abolition in a climate of suspicion and mistrust. George Kennan has proposed an immediate 50 per cent reduction across the board in the nuclear arsenals of the two superpowers.

For stability, however, the degree of asymmetry must be limited in that neither party may be permitted a preponderance sufficient to eliminate the effective capability of the other to inflict unacceptable damage, while still retaining his own. Such a preponderance may, paradoxically, be more easily obtainable with smaller than with larger numbers. Any process of gradual reduction must be very carefully planned if the capabilities of both sides are to remain at all times within a zone (in conceptual terms) of stable mutual deterrence. These points have been brought out mathematically by two Canadian analysts, Albert Legault and George Lindsey, and are illustrated in Figures 4.1 and 4.2 which, with the help of my Ministry of Defence and CCADD colleague Jim Barnes, have been adapted with permission from their work.[12]

In Figure 4.1 the horizontal axis is a measure of the number of missiles possessed by country X; the vertical axis is a measure of the number of missiles possessed by Y. U_x is the number of missiles which Y regards as sufficient for X to inflict unacceptable damage upon him; U_y is the number which X regards as sufficient for Y to cause him unacceptable damage. At point A each side possesses sufficient missiles

Figure 4.1: Stable and Unstable Deterrence

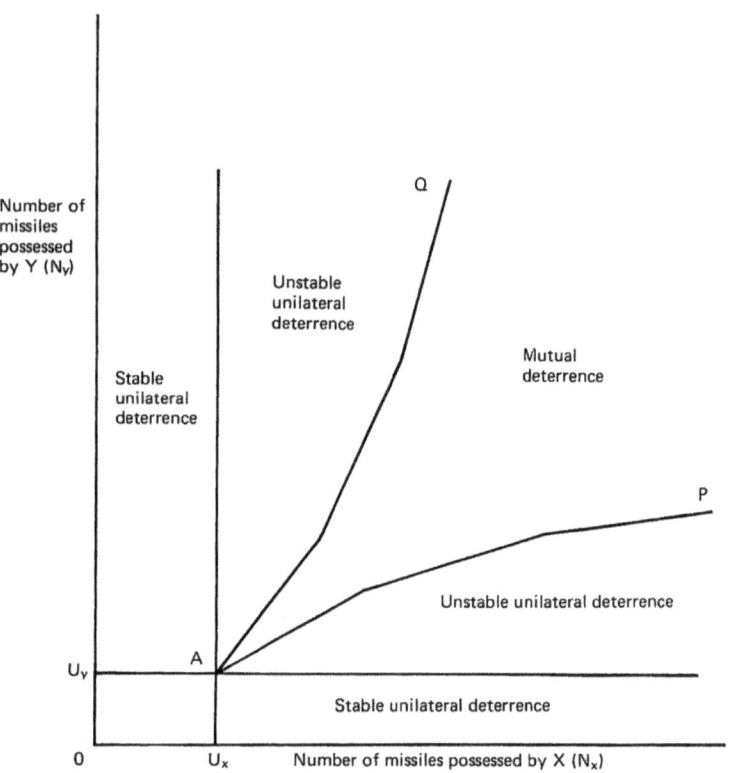

Source: Adapted from Albert Legault and George Lindsey, *The Dynamics of the Nuclear Balance* (Cornell University Press, Ithaca, NY, and London, 1974), Fig. 36, p. 182.

to be regarded by the other as capable of inflicting unacceptable damage. In terms of weapon numbers this is, for Legault and Lindsey, the lowest point of a relationship of mutual deterrence.

There are two regions of what they describe as 'stable unilateral deterrence'. In each of them only one side is perceived to have sufficient missiles to cause unacceptable damage to the other. The deterrence is unilateral because only the weaker is thereby deterred from launching his missiles; but in Legault and Lindsey's definition of 'stability' (the absence of rational motive to launch an attack) it is stable because the stronger need not fear unacceptable damage from the

In Defence of Deterrence

weaker and therefore has no incentive to launch a pre-emptive strike against him. Within the two zones of 'unstable unilateral deterrence' it is different. In the zone to the right of the line AP X has a clear numerical superiority, and although Y possesses sufficient missiles to inflict unacceptable damage upon X in a first strike, the retaliatory power of X is still more than enough to obliterate Y; therefore if Y is rational he is deterred. Nevertheless X may calculate that Y may commit an irrational act which X could prevent by making a first strike against Y's missiles. Y does not possess enough missiles to survive a first strike by X with a capability to inflict unacceptable damage upon X in retaliation — that is he does not possess an effective second-strike capability. Not only is X not deterred from launching a first strike, but he may have an incentive to do so; hence the description of this state of deterrence as 'unstable'. Within the zone to the left of the line AQ Y has a similar incentive and capability for a disarming first strike. Between the lines AP and AQ, however, there is a zone of mutual deterrence in which each side possesses sufficient missiles to survive a first strike from the other and to mount a retaliatory strike which would cause unacceptable damage to the aggressor.

It follows that if both sides possess massive nuclear armouries but start a process of gradual reduction from within the zone of mutual deterrence, they must remain within the boundaries AP and AQ unless one side is prepared to put itself at risk. Figure 4.2 illustrates the kind of step-by-step process which has to be undertaken and in which the risk becomes more acute as the numbers are reduced. At Step 6 X has abdicated his ability to inflict unacceptable damage upon Y, but is still exposed to the danger of such damage from Y. In terms of Figure 4.1 deterrence is now stable but unilateral. Had X moved to the left of AQ in Step 2 or Step 4, he would, although still possessing a capability to inflict unacceptable damage, have placed himself in a position where Y had an incentive to launch a disarming first strike against him, and would thus have moved into the zone of unstable unilateral deterrence. Moreover the shapes and positions of the boundaries AP and AQ are sensitive to the characteristics of the strategic weapon systems possessed by each side, including their vulnerability. Special care has to be exercised if one or both sides possess missiles with MIRVs (multiple independently-targeted re-entry vehicles). MIRVed missiles increase the striking power of the side possessing them and can cause the zone of mutual deterrence to become much narrower (and under some circumstances to disappear close to the point A on the diagrams).

The mathematical concept of stability propounded by Legault and

Figure 4.2: Paths towards Disarmament

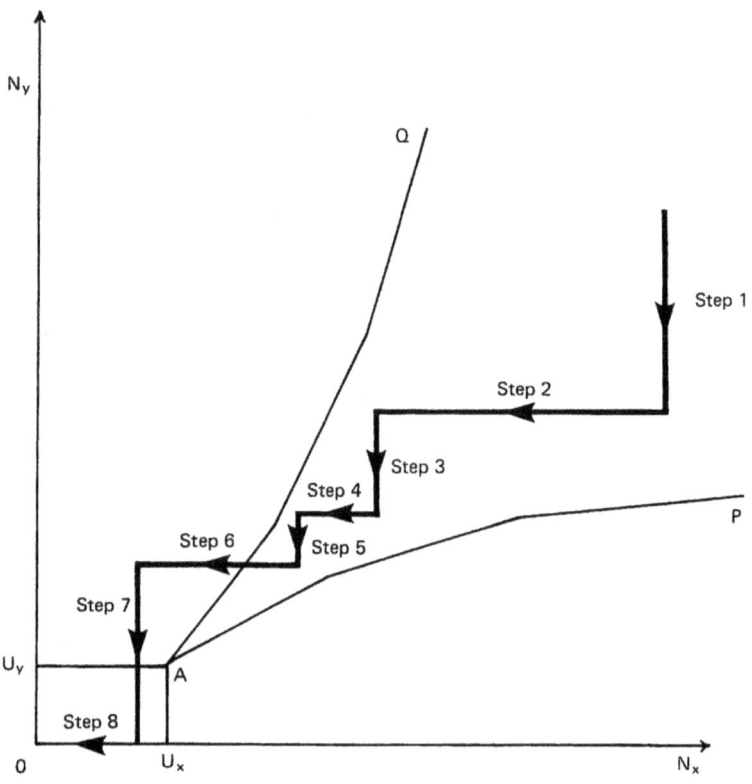

Source: Adapted from Albert Legault and George Lindsey, *The Dynamics of the Nuclear Balance* (Cornell University Press, Ithaca, NY, and London, 1974), Fig. 40, p. 192.

Lindsey also extends to a zone of 'no deterrence' within the rectangle OU_yAU_x on Figure 4.1, in which neither side regards the other as capable of inflicting 'unacceptable damage' and hence in their view has no incentive to eliminate the other's capability. However, entry into this zone, and equally the attainment of point O representing complete nuclear disarmament, is fraught with danger; first, because of the need for each side to be assured of the effective compliance of the other

In Defence of Deterrence

with the steps agreed for the reduction of weapon levels, and second, because of the risk, if point O were ostensibly reached, that one side might seek to gain a decisive advantage by the clandestine retention of a small number of weapons. In 1945 the Japanese found 'unacceptable' a degree of damage far short of the levels discussed earlier in this essay. In interpreting stability so narrowly in terms of the absence of a rational motive for a disarming first strike Legault and Lindsey seem to me not only to underestimate the difficulty of assessing capabilities with mathematical exactitude but also to take insufficient account of a latent political instability inherent in the psychological impact, at least upon democratic governments and the peoples to whom they are accountable, of the threat of even a few nuclear weapons to which no effective counter is apparent. They perhaps recognise this when they say that 'it is easy to see why a country not completely convinced of the good intentions of another will feel much safer when both have many weapons than when each has only a few'.[13] I would myself incline to the view that if two powers each had 100 delivery vehicles, or even one 100 and the other 50 (provided that neither could eliminate the retaliatory capability of the other while retaining a significant capability of its own), the balance between them would be decidedly more stable, and the risk of the political or military use of nuclear weapons less, than if one had three or four and the other none at all. To quote the Canadian analysts again, 'It should be clear that much more is involved in nuclear disarmament than the simple abolition of an unmitigated evil.'[14]

We may now turn to the ethical issues involved. Gordon Dunstan has pointed out in Chapter 2 that in relation to the world, its politics and its peace, Christian communities have been stretched between two convictions: the one that Christians should try to live as though Christ's rule on earth had already begun; the other that, so long as the Kingdom of God has not come on earth, the world must be governed under God's authority but within the terms of its own politics and jurisprudence – an earthly hope having become an eschatological hope that God's righteousness will triumph in the end. These two strains have been reflected, on the one hand, in the honourable tradition of Christian pacifism, and, on the other, in the equally honourable series of attempts to establish, in the context of the world as it is, a doctrine and set of rules governing the circumstances in which it may be justifiable to employ military force (*jus ad bellum*) and the constraints within which that force should be applied (*jus in bello*).

The case for unilateral nuclear disarmament within the tradition of

Christian pacifism is set out by Bruce Kent in Chapter 3. The present chapter is concerned with the ethical implications of an approach which recognises the existence of nuclear weapons, mistrusts the instability inherent in imbalances of the kind discussed above and seeks to engage the machinery of nuclear balance in the cause, shared with the pacifist or the nuclear disarmer, of preventing the use of nuclear weapons as instruments of either military or political aggression. Barrie Paskins has reminded us that morality inheres in the intention as well as the action;[15] and we must therefore examine in some depth the intentions underlying a strategy of nuclear deterrence.

A strategy of deterrence has two basic objectives; one is to prevent war or to terminate it at an early stage; the other is a political objective based upon perceptions of the geo-political scene. For the West this political objective is to ensure that we do not fall under the military dominance or the political hegemony of a country (the Soviet Union) under whose dominance we have no wish to fall and whose regime is of a type notable for the number of people who have been prepared to risk their lives to try to escape from it. It is sometimes asserted as self-evident that it is better to be red than dead. This is a bogus question; being red and being dead are not alternatives in the strict logical sense. Very large numbers of people have ended up dead under red regimes without being given any option. Conversely, the intention of the proponent of deterrence, as much as that of the pacifist, is to remain neither red nor dead; and although it may seem obvious from the comfortable distance of the British Isles that it is better to be red than dead, we should pause to consider the number of those whom closer acquaintance has impelled to make a different assessment. A Russian analyst might likewise express his political objective in terms of deterring yet another attack from the West inspired either by Yankee imperialism or by German revanchism, or possibly even in terms of deterring interference with the peaceful construction of socialism.

These basic objectives can be tested against the traditional criteria of *jus ad bellum* — lawful authority, just cause, and righteous intention[16] — and can be judged to meet them as adequately as the medieval conflicts with which the schoolmen were wrestling. More serious problems arise when a strategy of deterrence is tested against the requirement of *jus in bello* for proportion and discrimination. As to proportion, the good resulting from the deterrence of war, whether it be defined as peace, freedom or some other desirable political objective, is more than proportional to the morally neutral action of not fighting; but there is more room for argument about whether the military or political advantages derived from actually using military force can ever be proportional to the evil involved in doing so in a way which, either directly or as a more or less inevitable corollary, is indiscriminate as between combatants and non-combatants.

This point applies particularly, but not only, to a strategy in which nuclear weapons play a substantial part. Nuclear weapons are not the only weapons that can be used indiscriminately; it is probable that more people were killed in Hamburg or Dresden than in Hiroshima or Nagasaki, and McGeorge Bundy has reminded us that Hiroshima was construed as a highly military target and that the bomb exploded within 100 metres of its aiming point.[17] They are, however, distinguished by the amount of destruction that a single weapon can cause, which makes 'collateral damage' an almost inevitable corollary, except perhaps in virtually uninhabited regions or in certain specialised uses, as for example at sea. They are further distinguished by the inevitability of residual radiation. The effects of this are difficult to quantify; but a recent United Nations report has estimated that, for every kiloton yield of nuclear-fission explosions in the atmosphere, one person dies prematurely with cancer or genetic damage.[18] Nor, as pointed out earlier, should we be mesmerised by small numbers into treating what might seem comparatively minor nuclear explosions lightly.

A strategy of deterrence which includes a substantial nuclear element is criticised by Barrie Paskins in Chapter 5 on the ground that, even though it may have as a basic objective the prevention or termination of war, it nevertheless involves a conditional intention to use weapons whose effects are, other than in exceptional cases, virtually certain to kill non-combatants. Cardinal Hume argued in 1980 that 'it is quite clear from the authoritative teaching of the Second Vatican Council that the indiscriminate killing of civilian populations is immoral and can never be justified'.[19] Paskins would go on to say that to intend, even if only conditionally, to commit an immoral act is itself immoral; and that a strategy of nuclear deterrence (or a deterrent strategy involving any other element of mass destruction, such as conventional bombing or chemical attack against populations) is therefore morally unacceptable. In 1976, for example, the Roman Catholic Bishops of the United States declared that 'not only is it wrong to attack civilian populations but it is also wrong to threaten to attack them as part of a strategy of deterrence'.[20]

I do not, with respect, find the formulation of the American Roman Catholic Bishops very satisfactory in the terms stated. If it is derived directly from the Second Vatican Council, it may be reflecting a widely disseminated English translation of the relevant passage of *Gaudium et Spes* to the effect that 'all warfare which tends to the destruction of entire cities or wide areas with their inhabitants is a crime against God and man'. As suggested in a note to this essay,[21] the sense of the Latin appears to be more accurately conveyed in an alternative translation, in a different English edition of the documents of Vatican II, to the effect that 'any act of war *aimed indiscriminately* at the destruction of entire cities or of extensive areas along with their population is a crime against God and man himself' (my emphasis). Second, the term 'threaten' is imprecise in that a threat may inhere in a capability or in an intention or in some combination of the two, and may be either explicit or implicit. I therefore propose to follow Barrie Paskins in analysing the ethical problem in terms of a declaration of a conditional intention.

As thus reinterpreted, the Bishops' formulation raises the questions whether a strategy of nuclear deterrence really does depend upon a conditional intention aimed indiscriminately at the destruction of civilian populations and whether whatever conditional intention the strategy involves is indeed immoral, and if so to what degree. Much attention has been devoted in the United States, both within government and among independent analysts, to the development of strategic concepts which offer at least a range of options alternative to an all-out counter-offensive against cities and which may accordingly be regarded as a more credible, if less devastating, deterrent since they do not automatically invite a further all-out counter-city response and may therefore be more plausibly selected. The first official exposition of such a concept came from Secretary of Defense McNamara at Ann Arbor, Michigan, on 16 June 1962; the most recent, an analysis of what has become known as Presidential Directive no. 59, from Secretary of Defense Harold Brown to the United States Naval War College in Washington on 20 August 1980. Secretary Brown's address[22] was noteworthy for his emphasis that Presidential Directive no. 59 constituted neither a new strategic doctrine nor a first-strike strategy, and for his recognition that any notion of victory in nuclear war is unrealistic. In stressing the desirability of ensuring that the Soviet leadership entertained no illusions on this latter point, he said that the United States has never had 'a doctrine based simply and solely on reflexive, massive attacks on Soviet cities' and spoke of making clear to the Soviet leadership that the United States could 'exact an unacceptably high price in the things [they] appear to value most – political and military control, military force, both nuclear and conventional, and the industrial capability to sustain a war'. Similarly the British government has stated that its 'concept of deterrence is concerned essentially with posing a potential threat to key aspects of Soviet state power'.[23]

These formulations bring into consideration the principle of double effect, first adumbrated in the discussion by St Thomas Aquinas of killing in self-defence,[24] and progressively developed by subsequent Catholic theologians. The basis of the principle is that if an effect of an action is not directly intended by the actor, it should not be imputed to him morally in the same way as a directly intended effect. If therefore the indiscriminate killing of non-combatants (using the term in a broad sense) is not of the essence of the destruction of military forces or key aspects of state power, as it is not, then an intention to destroy military forces or key aspects of state power should not be judged morally as though it were aimed indiscriminately at the slaughter of civilians. If, however, the nature of the weapon used is such that some destruction of non-combatants is, while not of the essence, so nearly inevitable as to be a clearly foreseeable consequence, then it is a factor which must be taken into account in considering whether the projected action can be justified against the criterion of proportion.

Nevertheless, while the principle of double effect clearly brings into question whether the conditional intention expressed in a strategy of nuclear deterrence is to be condemned as being indiscriminately aimed at killing non-combatants, some may feel that the problem cannot be disposed of so easily. So let us analyse the intention further. What we are discussing is a *conditional* intention. Barrie Paskins has well brought out[25] that deterrence cannot be morally justified on the ground that I am bluffing and do not really mean it, for if I am bluffing my bluff can be all too easily called and its fragility as deterrence exposed. But this is only to say that, when I seek to deter by expressing the conditional intention 'if A, X', I must not have formed an actual intention not to do X. I do not have to form an actual intention to do X until the condition A is satisfied; what I am doing in deterrence is keeping my adversary persuaded that there is a measurable probability that if A I will do X, and the extent to which I can only do so by actually intending to do X is a matter of empiricism rather than of logic. However, as Ronald Hope-Jones points out in Chapter 8, this argument is valid for the President or the Prime Minister but can hardly be valid for the submarine commander whose part in deterrence must involve an actual intention to fire if he receives a properly authenticated order to do so from the proper authority. But this is not a matter of mindless obedience; he knows that the proximate condition for his firing is simply the final step in the much wider conditional process which I have described as 'if A, X'. And although the deterrent declaration 'if A, X' certainly involves for the submarine commander, and may involve for the President or the Prime Minister, an actual intention to commit, if certain conditions are fulfilled, an action of which the killing of non-combatants is a clearly foreseeable consequence, it seems reasonable to posit a moral distinction between a simple intention X and a conditional intention 'if A, X' when a principal objective of forming and declaring the conditional intention is to secure, so far as lies in one's power, that A shall not come to pass and that the condition shall not be fulfilled.

I have suggested that it is at least open to argument whether the conditional intention involved in a strategy of nuclear deterrence is a conditional intention aimed indiscriminately at killing non-combatants; whether the declaration of the conditional intention necessarily involves an actual intention; and whether, even if an actual intention has been formed in the terms 'if A, X', it is morally equivalent to a simple intention X. I do not believe, therefore, that a strategy of nuclear deterrence can be dismissed simply, as by Bruce Kent implicitly in Chapter 3 and Barrie Paskins explicitly in Chapter 5, on the basis of the conditional intention which it may express. On the other hand, I am not seeking to argue that the conditional intention is so morally laudable or morally neutral as to involve no element of evil. What I am arguing is that, although the conditional intention may contain an element of moral evil, a strategy of deterrence involving the conditional intention may be the most effective way of securing the twin objectives of preventing war and checking political aggression and may therefore be a morally acceptable price to pay to achieve those objectives — a question which Admiral of the Fleet Lord Hill-Norton has identified as central to the ethical consideration of deterrence.[26]

In a very thoughtful paper Allan Parrent has observed that in the Sermon on the Mount we are given material of fundamental importance in making judgements of moral value, that is to say what is good, and judgements of moral obligation, that is to say what is right.[27] In the same paper he referred to one Oxford philosopher of the inter-war years, Lord Lindsay of Birker; rather surprisingly he did not refer to another, Sir David Ross, one of whose books, *The Right and the Good*, epitomises in its title precisely the distinction that Parrent has noted between two categories of moral judgement.

Ross was a great Aristotelian scholar; and, like Aristotle arguing against the theoretical idealism of Plato, so when arguing against the theoretical utilitarianism of G.E. Moore he sought to relate to actual situations in which real people find themselves. In *The Right and the Good* he develops a theory of 'prima facie duties' (which might perhaps equally have been described as 'conditional duties'), by which he means acts which may at any given moment present themselves to a greater or lesser extent as duties, but which may also be compared in respect of other morally significant characteristics. He goes on to say:

> When I am in a situation, as perhaps I always am, in which more than one of these prima facie duties is incumbent on me, what I have to do is to study the situation as fully as I can until I form the considered opinion (it is never more) that in the circumstances one of them is more incumbent than any other; then I am bound to think that to do this prima facie duty is my duty *sans phrase* in the situation.[28]

I believe that a similar line of argument can be applied to our consideration of the morality of the conditional intention that underlies nuclear deterrence and can underlie other forms of deterrence also. Although to have such a conditional intention may not be *good*, and may even involve elements of evil, I suggest that it may be *right* if, by virtue of being in the circumstances the most likely means of securing that peace shall be preserved and that nuclear weapons shall not be used either by myself or by others, it is the least evil of a number of prima facie evil courses that may lie before me.

Like the canonists and moralists to whom Gordon Dunstan refers in Chapter 2, I have gone outside the strictly theological tradition into that of secular philosophy in order to arrive at the notion that the harbouring of a conditional intention involving elements of evil may be in the circumstances right, as the least evil of the courses open to me. I do not believe, however, that such a notion is incompatible with a Christian faith. The Christian message is one of hope and joy; but it is also part of that message that a price may have to be paid in order that the hope of joy may be fulfilled. I have no intention of decrying the honourable tradition of Christian pacifism, expounded by Bruce Kent in Chapter 3. My submission is that more than one view of these questions can legitimately be held within a framework of Christian belief; and that historical, political and strategic analysis suggests that prudent multilateral disarmament, backed by a stable though reducing balance of deterrence, offers a surer means than unilateral disarmament of maintaining both peace and the freedom underlying the values which, whatever their defects, most people in the West are anxious to preserve.

We must recognise that these arguments apply to deterrence in general and that, if common political objectives are assumed, there are no moral grounds for distinguishing British deterrence from American deterrence or NATO deterrence. There are, however, one or two observations that can be made with reference to such questions as whether Britain should have adopted the Trident ballistic-missile system as a successor to Polaris, or whether she should have offered bases for American cruise missiles. It is morally respectable (though in my view not necessarily right) to refuse to have anything to do with deterrence on the ground of the destruction that (even if only in conditional intention) we may wreak on others. It is less easy to construct a moral rather than a prudential argument (the moral argument is more one of our duty to our families and our fellow citizens) for refusing to have anything to do with a particular system of deterrence because we are afraid of what we might suffer at the hands of others; and even in prudence such a refusal runs counter to the historical evidence that nuclear weapons have been used only against a nation which had no means of comparable retaliation. And, to put it mildly, it is morally ambiguous to abjure on moral grounds British possession of nuclear weapons, or the stationing of American nuclear weapons in Britain, while being prepared to accept (or passively acquiesce in) the protection of a security system, founded in the North Atlantic Alliance, which rests in the last resort upon the strategic nuclear power of the United States.

But those who meet this last argument by condemning a NATO strategy of which nuclear deterrence is a significant component, and advocating the withdrawal of the United Kingdom from the Alliance, must face other questions. Nuclear weapons offer a means, however imperfect, of seeking to assure defence and deterrence relatively cheaply.[29] Those who reject nuclear deterrence on moral grounds must say whether, even if they accept that they may have laid themselves open to nuclear blackmail, they are prepared to support the conventional defence effort implicit in non-nuclear deterrence, with the corollary of a greater diversion of human and material resources from other potential applications; or whether they abjure deterrence altogether. And just as the statesman who commits his country to war, or the general who commits his army to battle, must weigh in his conscience the death to which he may be sending others, so must the pacifist weigh in *his* conscience, and be prepared to carry as *his* moral burden, the oppression and the injustice which others may suffer as a result of a policy which he has adopted on grounds which are for him more morally or spiritually compelling. While, however, my own submission that nuclear deterrence may be morally acceptable as being in certain circumstances right, even if not necessarily good, suggests that the West may legitimately deploy a strategic nuclear deterrent, it is not relevant to the question whether, as the government has consistently maintained,[30] Britain can most significantly contribute to Western deterrence through the provision of a Trident force, and do so without serious detriment to other military programmes; or whether, as a number of independent analysts have suggested,[31] the cost of the Trident programme cannot be reconciled with the notion that the United Kingdom should sustain a balanced contribution of conventional forces to the Atlantic Alliance.

The two Christian convictions identified by Gordon Dunstan in Chapter 2, and reflected in the tradition of Christian pacifism and in the search for a Christian definition of the just war, may be respectively characterised, in a world in which power has repeatedly been rampant, as concepts of abdication from power and management of power. Support for both these concepts can be derived from Scripture, though Dunstan pertinently reminds us that we must be scrupulous not to wrench particular passages from their exegetical context. We may also reflect, in the context of the management of power, that if the Franks had not defeated the Arabs at Tours in 732 we should perhaps have grown up in a Muslim culture and, as Gibbon put it, the interpretation of the Koran now be taught in the schools of Oxford.[32]

But, however much these two traditions may have diverged in the past, we must recognise that in the late twentieth century, when we have seen not only conventional warfare on an unprecedented scale but also the invention of nuclear weapons, the objectives of the Christian pacifist and of the Christian manager of power must be the same — the prevention of destruction and the preservation of peace. The questions at issue between them relate to the means of achieving those objectives and the nature of the price that must be paid. On the means, we may pause to ponder Shelley's lines:

And much I grieved to think how power and will
In opposition rule our mortal day —
And why God made irreconcilable
Good and the means of good.[33]

On the price, Michael Walzer has this to say:

Total destruction . . . is the danger that has faced mankind since 1945, and our understanding of nuclear deterrence must be worked out with reference to its scope and imminence. Supreme emergency has become a permanent condition. Deterrence is a way of coping with that condition, and though it is a bad way, there may well be no other that is practical in a world of sovereign and suspicious states. We threaten evil in order not to do it, and the doing of it would be so terrible that the threat seems in comparison to be morally defensible.[34]

Theodore Weber has suggested that the arrangements which human beings make or attempt to make by human means for securing their present and future are both futile and sinful.[35] It is as true as of any other human activities to say of attempts to preserve peace by the management of power and the deterrence of aggression that they partake of futility and sin. But I would submit in all humility that the efforts of decision-takers to tackle these questions with a proper sense of responsibility towards the issues are neither wholly futile nor (dare I say it?) wholly sinful.

APPENDIX: NUCLEAR WEAPONS AND PREVENTING WAR

1. Nuclear weapons have transformed our view of war. Though they have been used only twice, half a lifetime ago, the terrible experience of Hiroshima and Nagasaki must be always in our minds. But the scale of that horror makes it all the more necessary that revulsion be partnered by clear thinking. If it is not, we may find ourselves having to learn again, in the appalling school of practical experience, that abhorrence of war is no substitute for realistic plans to prevent it.
2. There can be opposing views about whether the world would be safer and more peaceful if nuclear weapons had never been invented. But that is academic; they cannot be disinvented. Our task now is to devise a system for living in peace and freedom while ensuring that nuclear weapons are never used, either to destroy or to blackmail.
3. Nuclear weapons are the dominant aspect of modern war potential. But they are not the only aspect we should fear. Save at the very end, World War II was fought entirely with what are comfortably called 'conventional' weapons, yet during its six years something like fifty million people were killed. Since 1945 'conventional' war has killed up to ten million more. The 'conventional' weapons with which any East-

West war would be fought today are much more powerful than those of 1939-1945; and chemical weapons are far more lethal than when they were last used widely, over sixty years ago. Action about nuclear weapons which left, or seemed to leave, the field free for non-nuclear war could be calamitous.

4. Moreover, whatever promises might have been given in peace, no alliance possessing nuclear weapons could be counted on to accept major non-nuclear defeat and conquest without using its nuclear power. Non-nuclear war between East and West is by far the likeliest road to nuclear war.

5. We must therefore seek to prevent any war, not just nuclear war, between East and West. And the part nuclear weapons have to play in this is made all the greater by the facts of military power. The combination of geography and totalitarian direction of resources gives the Soviet Union a massive preponderance in Europe. The Western democracies have enough economic strength to match the East, if their peoples so chose. But the cost to social and other aims would be huge, and the resulting forces would still not make our nuclear weapons unnecessary. No Western non-nuclear effort could keep us safe against one-sided Eastern nuclear power.

6. An enormous literature has sprung up around the concepts of deterrence in the nuclear age. Much of it seems remote and abstruse, and its apparent detachment often sounds repugnant. But though the idea of deterrence is old and looks simple, making it work effectively in today's world needs clear thought on complex issues. The central aim is to influence the calculations of anyone who might consider aggression; to influence them decisively; and crucially, to influence them before aggression is ever launched. It is not certain that any East-West conflict would rise to all-out nuclear war: escalation is a matter of human decision, not an inexorable scientific process. It is perfectly sensible — indeed essential — to make plans which could increase and exploit whatever chance there might be of ending war short of global catastrophe. But that chance will always be precarious, whether at the conventional or the nuclear level; amid the confusion, passions and irrationalities of war, escalation must always be a grave danger. The only safe course is outright prevention.

7. Planning deterrence means thinking through the possible reasoning of an adversary and the way in which alternative courses of action might appear to him in advance. It also means doing this in his terms, not in ours; and allowing for how he might think in future circumstances, not just in today's. In essence we seek to ensure that, what-

ever military aggression or political bullying a future Soviet leader might contemplate, he could not foresee any likely situation in which the West would be left with no realistic alternative to surrender.

8. Failure to recognise this complicated but crucial fact about deterrence — that it rests, like a chess master's strategy, on blocking off in advance a variety of possible moves in an opponent's mind — underlies many of the criticisms made of Western security policy. To make provision for having practical courses of action available in nuclear war (or for reducing its devastation in some degree by modest civil defence precautions) is not in the least to have a 'war-fighting strategy', or to plan for nuclear war as something expected or probable. It is, on the contrary, a necessary path to deterrence, to rendering nuclear war as improbable as we humanly can. The further evolution last year of United States nuclear planning illustrates the point. The reason for having available a wider range of 'non-city' target options was not in order to fight a limited nuclear war — the United States repeatedly stressed that it did not believe in any such notion — but to help ensure that even if an adversary believed in limited nuclear war (as Soviet writings sometimes suggest) he could not expect actually to win one.

9. The United Kingdom helped to develop NATO's deterrent strategy, and we are involved in its nuclear aspects at three main levels. First, we endorse it fully as helping to guarantee our security, and we share in the protection it gives all Alliance members. Second, we cooperate directly, like several other members, in the United States power which is the main component of the nuclear armoury, by making bases available and providing certain delivery systems to carry United States warheads. Third, we commit to the Alliance nuclear forces of various kinds — strategic and theatre — under our independent control. The details of all this are matters of debate, which the Government welcomes. But the debate should recognise that positions which seek to wash British hands of nuclear affairs, while continuing (as NATO membership implies) to welcome United States nuclear protection through the Alliance, offer neither moral merit nor greater safety. Whether we like the fact or not, and whether nuclear weapons are based here or not, our country's size and location make it militarily crucial to NATO and so an inevitable target in war. A 'nuclear-free' Britain would mean a weaker NATO, weaker deterrence, and more risk of war; and if war started we would if anything be more likely, not less, to come under nuclear attack.

10. The East-West peace has held so far for thirty-five years. This is a

In Defence of Deterrence

striking achievement, with political systems so sharply opposed and points of friction potentially so many. No-one can ever prove that deterrence centred on nuclear weapons has played a key part; but common sense suggests that it must have done. Deterrence can continue to hold, with growing stability as the two sides deepen their understanding of how the system must work and how dangers must be avoided. Not since the Soviet gamble over Cuba in 1962 have we come anywhere near the brink. It is entirely possible, if we plan wisely, to go on enjoying both peace and freedom — that is, to avoid the bogus choice of 'Red or dead'.

11. To recognise the success of deterrence is not to accept it as the last word in ensuring freedom from war. Any readiness by one nation to use nuclear weapons against another, even in self-defence, is terrible. No-one — especially from within the ethical traditions of the free world, with their special respect for individual life — can acquiesce comfortably in it as the basis of international peace for the rest of time. We have to seek unremittingly, through arms control and otherwise, for better ways of ordering the world. But the search may be a very long one. No safer system than deterrence is yet in view, and impatience would be a catastrophic guide in the search. To tear down the present structure, imperfect but effective, before a better one is firmly within our grasp would be an immensely dangerous and irresponsible act.

Source: Reproduced by permission from *Statement on the Defence Estimates 1981*, Cmnd 8212 (HMSO, London, 1981), vol. 1, pp. 13-14.

Notes

1. For example, an article by Otto von Gottberg in *Jungdeutschland-Post* on 25 January 1913 (quoted in Fritz Fischer, *War of Illusions* (Chatto and Windus, London, 1975), p. 193) acclaimed war as the 'noblest and most sacred manifestation of human activity'. The view of war as a 'process of purification' was found even in Church circles — see Fischer, *War of Illusions* (p. 253), for references to the *Allgemeine Evangelisch-Lutherische Kirchenzeitung*. And there is no reason to suppose that this was a uniquely German phenomenon.
2. Reproduced in Correlli Barnett, *The Swordbearers* (Eyre and Spottiswoode, London, 1963), facing p. 64.
3. John Terraine, *The Smoke and the Fire* (Sidgwick and Jackson, London, 1980), p. 44, see pp. 35-47 for an interesting discussion of 'The Great Casualty Myth'.
4. Quoted in Michael Howard, *The Franco-Prussian War* (Fontana edn, Collins, London, 1967), p. 56.
5. General André Beaufre, *Deterrence and Strategy* (Faber and Faber, London, 1965), p. 35.

92   In Defence of Deterrence

6. Statement of Secretary of Defense Robert S. McNamara before the Senate Armed Services Committee, 22 January 1968 (US Government Printing Office, Washington, 1968), p. 50.
7. *The Future United Kingdom Strategic Nuclear Deterrent Force* (Defence Open Government Document 80/23, Ministry of Defence, London, 1980), p. 13.
8. *Strategic Survey 1980-1981* (International Institute for Strategic Studies, London, 1981), p. 15.
9. See photographs in Rose Coombs, *Before Endeavours Fade* (Battle of Britain Prints International Ltd, London, 1976), pp. 59-61.
10. John Reason and Carwyn James, *The World of Rugby* (BBC Publications, London, 1979), pp. 164-5.
11. George F. Kennan, Acceptance Speech on receiving the Albert Einstein Peace Prize, Washington, 19 May 1981; *Guardian*, 25 May 1981.
12. Albert Legault and George Lindsey, *The Dynamics of the Nuclear Balance* (Cornell University Press, Ithaca, NY, and London, 1974), ch. VII, 'A Theoretical Model of Deterrence', pp. 166-99.
13. Ibid., p. 194.
14. Ibid., p. 194.
15. Barrie Paskins and Michael Dockrill, *The Ethics of War* (Duckworth, London, 1979), p. 237.
16. St Thomas Aquinas, *Summa Theologiae*, 2-2, 40, 1; cited in F.H. Russell, *The Just War in the Middle Ages* (paperback edn, Cambridge University Press, 1977), p. 268.
17. McGeorge Bundy, 'Strategic Deterrence Thirty Years Later: What Has Changed?' in *The Future of Strategic Deterrence*, Part I, Adelphi Paper no. 160 (International Institute for Strategic Studies, London, 1980), p. 6.
18. United Nations document A/35/392, 12 September 1980, para. 260.
19. Cardinal Basil Hume, Address to the Convention of the World Disarmament Campaign, 12 April 1980.
20. National Conference of Roman Catholic Bishops, *To Live in Christ Jesus: A Pastoral Reflection on the Moral Life* (US Catholic Conference, Washington, 1976), p. 34.
21. The relevant passage of Article 80 of the Vatican Council Constitution *Gaudium et Spes* reads as follows:

> Omnis actio bellica quae in urbium integrarum vel amplarum regionum cum earum incolis destructionem indiscriminatim tendit, est crimen contra Deum et ipsum hominem.

This is rendered in a translation published by the Catholic Truth Society (London, 1966) as:

> All warfare which tends to the destruction of entire cities or wide areas with their inhabitants is a crime against God and man.

But this translation renders *actio bellica* by the general term 'warfare', rather than its more precise meaning 'act of war'; ignores the adverb *indiscriminatim* qualifying *tendit*; and leaves the verb *tendit in* in its original ambiguity without attempting to construe it in the sense either of 'extending towards' or 'being intended towards'. It does not seem unreasonable to turn for guidance to the Italian text, which reads as follows:

> Ogni atto di guerra che indiscriminatamente mira alla distruzione di intere città o di vaste regioni e dei loro abitanti, é delitto contro Dio e contro la stessa umanità.

In Defence of Deterrence 93

The verb *mira* has a precise connotation of 'aiming towards', suggesting that *tendit in* should be construed as a verb of aim or intention, qualified as in the Italian by the adverb *indiscriminatim*. The preferable English translation would therefore appear to be that of W.M. Abbott in his edition of *The Documents of Vatican II* (Geoffrey Chapman, London, 1966), which reads as follows:

> Any act of war aimed indiscriminately at the destruction of entire cities or of extensive areas along with their population is a crime against God and man himself.

22. Harold Brown, *The Objective of US Strategic Forces*: An Address to the Naval War College in Washington, 20 August 1980 (International Communication Agency, US Embassy, London, 1980).
23. *Future United Kingdom Strategic Nuclear Deterrent Force*, p. 6.
24. St Thomas Aquinas, *Summa Theologiae*, 2-2, 64, 7.
25. Paskins and Dockrill, *Ethics of War*, p. 69.
26. Admiral of the Fleet Lord Hill-Norton, 'After Polaris', *The Economist* (15 September 1979), p. 28.
27. Allan M. Parrent, 'The Sermon on the Mount: A Theology of Reconciliation and International Politics', paper delivered to the international conference of the Council on Christian Approaches to Defence and Disarmament, Friedewald, West Germany, 1980.
28. W.D. Ross, *The Right and the Good* (Clarendon Press, Oxford, 1930), p. 19.
29. See, for example, Defence Secretary John Nott's speech at Nottingham, 5 June 1981; or the analysis of defence resources by major programmes in *Statement on the Defence Estimates 1981*, Cmnd 8212 (HMSO, London, 1981), vol. 1, p. 66.
30. The White Paper published on 25 June 1981, *The United Kingdom Defence Programme: The Way Forward*, Cmnd 8288 (HMSO, London, 1981), para. 10, is the most recent example at the time of writing.
31. See, for example, David Greenwood, *The Polaris Successor System: At What Cost?* (Aberdeen Studies in Defence Economics no. 16, Aberdeen, 1980), p. 29.
32. Edward Gibbon, *History of the Decline and Fall of the Roman Empire*, ch. 52.
33. Percy Bysshe Shelley, *The Triumph of Life* (fragment published posthumously, 1824), ll. 228-31.
34. Michael Walzer, *Just and Unjust Wars* (Allen Lane, London, 1978), p. 274.
35. Theodore R. Weber, 'Theological Perspectives on Security, International Responsibility and Reconciliation', paper delivered to the international conference of the Council on Christian Approaches to Defence and Disarmament, Maryknoll, NY, 1978.

5 DEEP CUTS ARE MORALLY IMPERATIVE
Barrie Paskins

All-out nuclear war between the Soviet Union and the West would be immoral for two main reasons. First, it would visit immoral destruction upon non-combatants. Direct attacks on non-combatants would be part of any all-out nuclear war between the superpowers and long-term fall-out, raining down upon the earth for years after the holocaust, would bring agonising death to people throughout the world wholly innocent of involvement with the quarrels of the great powers. Second, the evil that all-out nuclear war would entail for all mankind would be out of all proportion to any good that might be secured. Anyone who replies 'Better dead than red' is inviting us to believe that some good or some lesser evil can be found which will outweigh the certain death of hundreds of millions of people, the very likely destruction of anything worth calling civilisation, genetic damage to future human generations, and possibly mutations among other species resulting in the extinction of man and/or damage to the atmosphere sufficient to destroy all green plants and all animals. 'Better dead than red' applied to all-out nuclear war is monstrous hubris. Who are we, beings with a life expectancy of decades, to discuss the entire future of the planet in terms deriving exclusively from our concerns, concerns which may be expected to count for nothing in a few millennia with or without nuclear war? There is megalomania, madness, in supposing that we can be important enough to make the judgement 'Better dead than red', given the practical meaning of that judgement relative to all-out nuclear war.[1]

One main purpose of deterrence as practised by the Soviet Union and by the West is to prevent all-out nuclear war. Is this deterrence morally legitimate? If not, what is to be done? To answer these questions, one has to begin by identifying the general character of the deterrents in question. The Soviet deterrent is thought to be somewhat different from the Western. Soviet planners appear to intend the massive early use of nuclear weapons large and small against a very wide range of targets in any major European war. This intention should not necessarily be interpreted as aggressiveness. The USSR has an immensely long border along which it confronts many enemies. It hopes that overall strength will deter attacks but is convinced that it has good reason to fear the attrition of war on several fronts and, as part of its offensive strategy for defence, looks to seize the military initiative as the way of winning any battle quickly. Nuclear weapons are seen as contributing to overall strength, and thereby to deterrence.

American and NATO strategy is to keep the peace by planning for military missions of increasing ('escalating') destructiveness. At 'the bottom of the escalation ladder' are non-nuclear ('conventional') forces designed to bear the brunt of any attack in Europe by Warsaw Pact forces which according to many indices outnumber the Western forces considerably. If the conventional forces could not hold and diplomatic efforts failed to halt the incursion, NATO has the capacity and reserves the right to be the first to use nuclear weapons. NATO distinguishes several distinct missions involving the use of nuclear weapons. The distinctions have varied from time to time and are to some extent debated. It is over-simplifying but not, I trust, misleading to distinguish as follows: (1) battlefield missions, with nuclear weapons fired from Europe or the adjacent sea, aimed at targets on a European battlefield not including Soviet territory; (2) theatre missions, with nuclear weapons fired from Europe or the adjacent sea, aimed at targets which may be inside the USSR; (3) strategic counter-force missions, fired from the US or the adjacent sea or the deep oceans, aimed at military targets in the USSR; (4) strategic counter-city missions, fired from the US, etc., aimed at Soviet cities.

From a Western viewpoint, escalation in the order (1) — (4) clearly represents nuclear war of growing intensity, with the bombardment spreading at each step to engulf the USSR (2), the US (3), and the non-combatant inhabitants of cities (4). Yet all commentators agree[2] that the prospects are poor of fighting any major war in Europe which does not degenerate ('escalate') into all-out nuclear war. Among the reasons for this are NATO's dependence on nuclear weapons to eke out its (real or supposed) conventional inferiority, and the Soviet determination to use nuclear weapons of all types early in any war. The prospects of limiting war in Europe are, all agree, still worse if the nuclear firebreak is crossed. For the scale of devastation would then be so great as to stir destructive passions incompatible with the cool, poker-player calmness that graduated escalation requires from both sides; and it is much easier to know that the enemy is firing nuclear weapons at one than to indulge in nice judgements about the ranking of these weapons in the enemy's escalation ladder.

Why, given these acknowledged difficulties, does NATO allocate scarce resources to maintaining the capacity for nuclear war at several different levels of intensity? The main reason is to be found in the delicate balance theory of deterrence. According to this theory, each side in the East-West confrontation is constantly reassessing what it can get away with in Europe and in the world. To take an important example, one need not be anti-Russian to suppose that the Soviet leadership has under constant review the question of whether, Western nerve and military might being what it appears in Moscow to be, Russia can with impunity occupy West Berlin, the city which has been a thorn in their flesh for decades, a show-place of Western freedom and affluence through which Warsaw Pact subjects have fled in their humiliating thousands. The danger, according to the delicate balance theory of deterrence, is that in the case of West Berlin or some other case one side or the other will miscalculate what it can get away with. To continue the example, if the Western plan for nuclear weapons looked solely to inter-continental bombardment of cities, then Soviet decision-makers might calculate that an American President would not dare to resist their seizure of West Berlin. The city is surrounded and indefensible; conventional forces counter-attacking elsewhere would be outnumbered; strategic attacks on the USSR would bring devastation on American cities. The Soviet leaders might under these circumstances count on the Americans giving in. A free city would be lost or all-out nuclear war would engulf the world.

The conventional forces together with the battlefield and theatre (and in some views the counter-force strategic) nuclear forces are intended to solve this problem so far as it can be solved. So long as the President has options other than suicide or surrender, it is argued, Soviet decision-makers cannot rationally count on a walk-over. With the conventional forces to fight a holding operation while conflict resolution is tried, and with the possibility of resort to nuclear war which may be capable of being kept limited, the President has other options. The city is saved and war prevented.

A second reason for NATO's multi-layered nuclear planning is summed up in the ugly but convenient term 'intra-war deterrence'. It is often said in popular debate that the strategy of deterrence will have failed if war breaks out in Europe. This over-simplifies to the point of misrepresentation. Military planners are charged to plan for (among other things) the worst; if nuclear war does break out in Europe, can nothing be done to keep it limited? Although all agree that the prospects for limitation are poor, some would argue that limitation is worth seeking. Other things being equal, the idea of striving for some measure of damage-limiting deterrence even within war seems to be common sense: it might work. Hence the importance of intra-war deterrence.

Deep Cuts are Morally Imperative

The Soviet and Western deterrents are, we have seen, somewhat different in structure and plan. The Soviet Union counts nuclear weapons as part of its overall strength, to which it looks for the deterrence of war and which it plans to use mightily in order to seize the offensive should war come. The West, too, values overall strength, but pictures deterrence as delicate and takes the credibility problem seriously, aiming to solve the problem by sustaining a capacity for graduated escalation.

Despite their differences, the Soviet and Western deterrents have much in common. For our present purpose the most important similarity is this: each side has the conditional intention to use nuclear weapons large and small against a wide range of targets including cities. This commitment to nuclear weapons is not a bluff: political will is tested remorselessly and any bluff will be called; weapons systems which are a bluff will be found out; and the personnel manning the deterrent have a resolution to carry out their orders which is no bluff. Military deterrence in general is the intention to use military force if the adversary oversteps certain limits. The Soviet Union intends to use nuclear weapons early. We in the West reserve the right to use nuclear weapons first and intend to use, if it comes to it, conventional weapons then nuclear weapons in increasingly destructive missions. Both East and West rely for the avoidance of war and of all-out nuclear war upon the conditional intention to wage war and all-out nuclear war.

If all-out nuclear war would be immoral, what of this conditional intention? Its aims — to keep the peace, to prevent aggression — are laudable. Furthermore, the West's conditional intention has the complexity it has, embodies the costly escalation ladder, in order to minimise the likelihood of miscalculation and thus, so far as possible, to avoid the intention becoming action and to limit the disaster should war come. Soviet offensive strength, too, doubtless has among its objectives to minimise the danger of miscalculation by Russia's enemies. On the other hand, all agree that if war in Europe breaks out then we are very likely to find ourselves engaged in all-out nuclear war. Normally, one thinks that if an action is immoral then the intention to carry out that action is also immoral. Not only highway robbery but also the robber caught before he can effect his intention is rightly punishable. Do the laudable aim and well-intentioned complexity or strength of deterrence soften the adverse judgement one would normally make? I am unable to believe that they do but the question is controversial, not least among the contributors to this book, and the underlying issues are the subject of intricate debate among philosophers. In this essay I confine my discussion of the matter to a terse summary of and comment on what I take to be the most important points advanced by those who consider deterrence morally defensible.

It is often urged that the conditional intention to wage all-out nuclear war is not as immoral as the action. The words 'as immoral as' are no part of our moral vocabulary and their meaning needs explanation which I have not found in the literature. If I plan to beat you up and do so, then there is more for me to reproach myself about than if my plan is thwarted. This indicates one thing that might be meant by holding that the intention to beat you up is 'less immoral than' the deed and could be applied to any immoral deed *considered in retrospect*. But retrospection is not the issue. The question is whether looking forward, deliberating what to do, one can in good faith ask whether a conditional intention is as immoral as the act intended. The answer is, surely, that the question, if clearly envisaged, is bound to be disingenuous. Prohibiting an action sets limits to what may be purposed, selected, intended, done; separate prohibition of intention is otiose. I doubt, therefore, whether deterrence can be defended by claiming that it is 'less immoral than' all-out nuclear war.

Sometimes, those who judge deterrence morally defensible deny that it can be equated with the conditional intention to wage, *in extremis*, all-out nuclear war. They urge one or more of three reasons for rejecting the equation. First, they point out that the decision to wage war lies in the future, a future we hope to avoid. What exists now, they urge, is the capability not the intention. Second, they emphasise that deterrence works by creating uncertainties in the mind of the opponent. Not our conditional intention to vaporise Moscow but Soviet uncertainty of whether we will do so is the mechanism of deterrence: their uncertainty not our intention. In reply to these two points, I grant that the decision to fire lies in a future we hope to avoid and that the measures we take operate via the hearts and minds of our opponents. But deterrence now is more than capability. The demonstration of political will is a necessary part of deterrence now, as defenders of deterrence will readily agree because without it the opponent's uncertainties may atrophy. What significance attaches to demonstrations of political will? In public our leaders insist on their resolve to use nuclear weapons if necessary. By speaking in earnest of necessity, they give the clearest possible public evidence that the West stands prepared *now* to envisage circumstances in which nuclear weapons will be used. In secret, I presume, newcomers to high office are briefed about the careful thought that has been given to the circumstances in which it

Deep Cuts are Morally Imperative

will be 'necessary' to use nuclear weapons. The decision to fire, if it is ever made, will surely not be taken by the government of the day with no reference to established thinking on the subject. As a citizen privy to no secrets whatever, I cannot but believe that the guide-lines for a briefing exist *now*, if only to minimise the danger of maverick decisions by some future government of the day. Taken together, the public pronouncements which are there for all to see and the secret positions which must exist amount to a conditional intention.

A third reason for refusing to equate deterrence and the conditional intention is that deterrence involves an actual intention to avoid having to give effect to the conditional intention. This actual intention shows that it is one-sided to equate deterrence and the conditional intention. My reply to this powerful argument is that the aim of avoiding war and preventing aggression, of averting the kind of occurrence that might occasion ('necessitate') our resort to nuclear weapons is a laudable aim, but too far from being within our power to secure to be described as an intention of ours. It is indeed the (main and most laudable) motivation of our deterrent activity, but intention and motive are different things. It is what we hope to achieve by means of deterrence, but hopes and intentions are different things. What I *intend* to do is what I judge, correctly or incorrectly, to lie within my power to do in the furtherance of often mixed motives. What I hope thereby to achieve is a further good, beyond my unaided power. The conditional intention to wage nuclear war is properly described in terms of intention, because the waging of nuclear war lies within our power; the prevention of war and aggression are goods we hope to achieve by means of the conditional intention. It is not, therefore, one-sided to equate deterrence and the conditional intention; though, of course, reference must be made, as I made it earlier, to the laudable aims and hopes of deterrence.

I conclude that the Soviet and Western deterrents are rightly characterised in terms of the conditional intention to wage, *in extremis*, all-out nuclear war; and that they as well as all-out nuclear war are immoral.

So what is to be done? Even in private life it is not always right to stop doing something at once on recognising its immorality. An example may make this plain. Consider a man conducting two adulterous affairs who comes to recognise that each of the two affairs is immoral. Suppose one affair can be ended coolly with little hurt — Thanks For The Memory — whereas the other mistress is suicidally dependent on the relationship. Plainly, it would be as wrong to terminate the latter affair at once as it would be to continue the former. The ending of our conditional intention to wage all-out nuclear war seems to involve the complexities of the more difficult of the two liaisons, not least because there are no God-given rules of disengagement. Hence all-too-fallible prudence as well as moral principle must guide us in withdrawing from our morally untenable position. The remainder of this essay seeks to identify the most prudent course of action available to us.

This is a large subject and I shall concentrate on only one aspect of it, namely the current controversy between unilateralists and their opponents. This focus needs, perhaps, little apology. A great deal has been said recently for and against unilateralism, and the argument manifestly stands in need of critical sifting. I shall argue for a type of Western unilateralism[3] — deep unilateral cuts in the West's deterrent — urging that the unilateralists must and can answer strategic questions. At the same time, I express considerable disquiet about the unilateralists' lack of interest in strategy.

If deterrence involving nuclear weapons is immoral then all should abandon it — the Soviet Union and China as well as the Western powers. There are two main strands of thought about how this universal withdrawal from a morally untenable position is to be secured.[4] One is to concentrate on multilateral arms control agreements, insisting that substantial reductions in the deterrent forces embody some sort of balance between East and West, and accepting that some arms control agreements may even legitimise increases in certain categories of weaponry.[5] Let us call this the multilateral approach. The other approach seeks to initiate a process of progressively decreasing tension and progressively drastic disarmament by means of substantial unilateral cuts by one side that are not made conditional upon specified responses by the other. Let us call this the unilateral approach. Unilateralists are not against multilateral agreements; rather, they look for further initiatives over-and-above multilateral agreements. Multilateralists are not necessarily against all unilateral initiatives, but they do insist on the overriding importance of a framework of mutuality. To formulate clearly and fairly the divergence between these two approaches is vital if we are to make a rational choice of the most prudent way forward. It is also difficult because one is not comparing like with like.

The multilateral argument has found widespread support among Western governments for decades. It has been debated, refined and complicated in official and unofficial forums. Arms control is the 'profession' of a good many civilians as well as being the major concern of important government and military departments. As Roy Dean makes plain in Chapter 6, it has gone through the fire of East-West negotiation and so has a place in political history as well as in the history of political ideas. In studying it one is therefore faced with the articulateness and sophistication that flow from the political commitment of successive governments together with, some would add, a sobriety that comes only from the responsibilities of office. The unilateral argument has not (yet) found support among governments. Its political life has been in the letter pages of newspapers and the heads of anxious citizens. It has not undergone institutional refinement and is an untried quantity. In trying to compare it with multilateralism, one is faced with a body of argument that is inevitably crude. Its rawness flows from its not (yet) having made its way into the corridors of power. A certain zest may well derive from the same source. With the debate between unilateralist and multilateralist thus unbalanced, any rendering of the principle issues is bound to be speculative. My aim is to disregard superficial polemic in search of the strongest argument that can be found on each side. This is a risky procedure because on both sides there are appalling arguments and in human affairs it is always possible for the worse argument to prevail over the better. But I can see no better way to seek the prudent course than by concentrating on the best arguments on either side.

The best argument for unilateralism known to me is, in outline, as follows. (1) The conflict, tension, fear and hatred between East and West is in part an underlying reality which existing levels of weaponry reflect and do not cause. But the possibility cannot be ruled out that the existing military confrontation has a certain life of its own, aggravates to a considerable, if unquantifiable, extent the poor state of East-West relations. As an ominous factor which decision-makers are bound to take into account and as a focus of competition which is unavoidable as things stand and contributory to ever-deepening mistrust as the competition intensifies, this military confrontation is a legitimate focus of concern among those who would reduce to the unavoidable minimum that which poisons East-West relations. (2) The problem is to initiate a process of disarmament which shall at least remove this needless aggravation of the bad relations between East and West. (3) In the search for a solution to this problem, we in the relatively free West bear a special responsibility because public opinion can have an effect on policy here, which is impossible in the closed society of the USSR. (4) Western initiatives aimed at starting a process of disarmament may fail, but they are worth trying because there is a reasonable chance of their succeeding and because the alternative multilateral approach seems most unlikely to extract East and West in the foreseeable future from the morally impossible position of conditionally intending to wage all-out nuclear war. (5) Such initiatives are therefore prudent and morally imperative.

This argument raises numerous questions for the unconvinced. Let us consider some (by no means all) of the matters arising in relation to parts (1)-(3) of the argument.

Thus (1), as I have stated it, is unlikely to arouse much disagreement from multilateralists, who usually agree that arms levels are far higher than they need to be but differ from unilateralists as to the means by which reductions are to be sought. Some unilateralists may consider that my formulation of (1) fails to go far enough by neither condemning existing arms levels as a *principal* source of tension nor even declaring categorically that they are a source of tension. I make no apology about this: I do not believe anyone knows more than that there is good reason to suppose that existing force levels aggravate East-West tensions to a considerable but unmeasurable extent.

Step (2) invites the multilateralist objection that other objectives including deterrence and stability are as important as, or more important than, the questionable adventure of disarmament. I comment on this extremely important criticism later.

Step (3) invites challenge from two sides. The multilateralist will object that Western public opinion in its very freedom can threaten the strategic balance. The backing this public opinion gave to isolationism and appeasement in the 1930s was an important factor in Hitler's miscalculation that he could dominate Europe by unresisted blitzkrieg, and citizens who support unilateralism now from a simplistic understanding of their civic duty are repeating the 1930s mistake of ignoring strategy. I agree with this objection and argue below that the unilateralists must *and can* face the strategic issues. More surprising, perhaps, than this multilateralist objection is the view espoused by some unilateralists. Professor E.P. Thompson has contended, if I understand him correctly, that mass movements in West *and East* are to initiate the desired process of disarmament. In arguing thus he seems to deny that we in the West have a special responsibility consequent upon our relative freedom. I find his position on this puzzling both because his view is not accepted by admirable Russians who should be in a position to judge[6] and because, paradoxically, his insistence that mass movements from West *and East* shall initiate disarmament reduces the urgency of making a start here and now.

The most formidable multilateralist criticism of the argument

Deep Cuts are Morally Imperative

(1)-(5) is that it ignores 'strategic realities'. I agree that the argument needs to be supplemented by strategic thought, but believe that the unilateralist can make out a much better strategic case than the multilateralist supposes. Because this aspect of the matter is largely ignored in current polemics, I shall touch on it at some length.

As we noted above, Western orthodoxy is that deterrence is *delicate*. Each side is constantly reassessing what it can get away with and constantly probing the will and capability of the opponent. There is a real danger of one side or the other deciding, on the strength of its probing, that some adventure is worth risking which, embarked upon, results in major damage to the other side's legitimate interests or in major war.

But it is not self-evident that deterrence is delicate. One might argue, and I think unilateralists should argue, that deterrence is robust rather than delicate. The argument is as follows. The dangers of miscalculation in Europe are so great, and are known on all sides to be so great, that no one dare take the risk. Antagonistic activity by East and West will be confined to actions which are not even remotely likely to result in European war. Deterrence will not prevent Soviet intervention in Afghanistan or Eastern Europe, American intervention in Latin America or South East Asia because the danger of European war is what underlies deterrence, and these things are not even remotely likely to result in European war. Neither side will dare to attempt anything, for example seizing West Berlin, remotely likely to result in war in Europe. Military forces are therefore prevented by deterrence from being a usable interest against the vital interests of either side. Even substantial Soviet preponderance in the military balance poses no real threat because Soviet leaders would not dare to run the risk, small though it may be, of precipitating a European war which all regard as likely to degenerate rapidly into nuclear conflagration. Deterrence does not involve a delicate balance because even indelicate imbalances cannot be exploited to pose any serious threat. Deterrence is robust rather than delicate.

The idea that deterrence is robust is very important for the unilateralist argument outlined above, for it explains how a Western initiative of deep cuts can be made *without undermining deterrence*. It answers the multilateralists on their strongest ground by challenging their fundamental contention that any major initiative would be bound to threaten the security system which, they claim, has prevented war between East and West for thirty years. But is it reasonable to contend that deterrence is robust? I shall outline and examine seven arguments against the robustness idea.

(A) It might be protested that unilateralism represents a failure of patience. The Western deterrent and such East-West accommodations as have been achieved are the product, it might be urged, of decades of patient endeavour, of cautious experiments some but not all of which have failed, of thoughtful correction of mistakes made in the going. All this the unilateralist is prepared to sacrifice because he lacks the patience to continue the slow but solid process of recent decades.

There is, I think, no doubt that much patience and ingenuity have gone to the making of the strategic *status quo* and some of the achievements have been valuable and remarkable. For example, the Partial Test Ban Treaty has saved tens of thousands from dying of cancers caused by fall-out and the Anti-Ballistic Missile Treaty has saved us from a vastly expensive and futile race to erect defences against inter-continental ballistic missiles. But there is a deep question which the unilateralist is entitled to press upon the multilateralist. Is it really credible that existing strategy and arms control is aimed at the morally imperative goal of abandoning the conditional intention to wage nuclear war? Is it not much more realistic to believe that strategists and arms controllers are working to prevent deterrence (including nuclear weapons) from becoming needlessly dangerous or expensive? Consideration of the objectives of strategists and arms controllers suggests to me that the elimination of nuclear weapons from their present vital, central role in deterrence is simply not among their aims. 'For the foreseeable future' they expect nuclear weapons, and the conditional intention to use nuclear weapons, to remain a vital part of strategy. One is scarcely being impatient if one jibs at the implicit readiness to continue with this immoral conditional intention for an unlimited future. The urge to find a way forward is not a failure of patience but a practical implicacation of the recognition that our present stance is morally untenable and current orthodoxy is prepared for this untenable position to continue forever.

(B) One might object to the view that deterrence is robust on the ground that such a view is reckless. Caution is the best policy in military affairs and caution requires us to adopt the working assumption that deterrence is delicate rather than to gamble on its being robust. This objection is distinct from (A) because it concerns safety, security. Even if principle requires more urgency, one can still question whether principled urgency is safe. This objection raises large issues which I cannot treat adequately in this essay but I shall attempt a three-part reply, arguing that: (a) deterrence as we know it is doubtfully safe in its own terms; (b) deterrence as we know it may well be incoherent;

Deep Cuts are Morally Imperative

and (c) it is question-begging to dismiss as incautious the view that deterrence is robust.

(a) Is deterrence as we know it safe? During the Cuba missiles crisis President Kennedy put the likelihood of war at between a half and a third. No one supposes that deterrence in the future will preclude crises, and there is certainly no reason in orthodox strategy to count on these crises being less risky than that of 1962. A half or third chance of nuclear war, endemic in the deterrent as normally conceived, is hardly safety.

(b) Is deterrence coherent? Let us look again at the credibility problem. Orthodox strategy would have us accept the following. For practical purposes we should picture the opponent as one who is prepared to risk nuclear war to (say) seize West Berlin providing that the probability of nuclear war is fairly low (but not negligible), but who will be deterred from (say) seizing West Berlin providing that the probability of nuclear war is raised somewhat (but not to certainty). It is not required that we suppose this to be true of the USSR; only that we accept that this is the prudent, cautious view.

I have no quarrel with the deeply pessimistic conception of international relations that underlies the argument. It would be recklessly naive to forget that the great powers throughout history have been cruel risk-takers, calculating their own advantage and seeking to maximise it with whatever seem the likeliest means to hand. But I am strongly disposed to think that the credibility problem, on which so much turns, is either insoluble or unreal. Consider: Soviet leaders must be presumed to have a fair knowledge of Western strategic writing. They therefore must know that Western opinion regards the prospect of keeping major war limited as poor. This must colour their estimate of the American President's readiness to go to war in defence of Europe. They must be as well aware as we are in the West that if the suicide-or-surrender dilemma was ever real then it still is because the prospects of graduated escalation working as planned are acknowledged to be poor, with the corollary that ordering even a bit of escalation for the sake of Europe is already very probably countenancing the destruction of American cities.

I do not say that this is what the Russians are thinking. My point is that we must consider the impact of this on the standard picture of the opponent which, we are told, a prudent attitude to deterrence requires. Let us call the opponent X. X is supposed to be reckless enough to risk all-out nuclear war to seize Berlin, providing he reckons that the likelihood of all-out nuclear war is pretty low. But at the same time, this reckless X is supposed to be deterred by graduated deterrence. Even though, as we have just seen, he is bound to regard graduated deterrence as failing to remove the Americans' suicide-or-surrender dilemma, he is nevertheless supposed to think that graduated deterrence raises the probability of Western reprisal sufficiently to make it unwise to seize Berlin. He is reckless; he is prepared to risk nuclear war providing the probability of it is pretty low; he imputes a fairly small extra probability to nuclear war if the West has graduated deterrence; and with this small increase in probability he is not prepared to risk nuclear war.

Is X believable? Of course, we do not *know* how the USSR deliberates and any fancy is conceivable. But once one formulates clearly what X is supposed to be like he emerges, I suggest, as quite incredible, wholly artificial. This is a drastic claim for a short essay. But the unilateralist should, surely, be raising fundamental questions. And here, I submit, is a fundamental question to be faced. A state reckless enough to risk nuclear war on the obviously fallible calculation that an alliance which is firmly integrated and politically strong will fail is not going to be deterred by marginal increments in the probability of nuclear war, is it? The question needs much further sifting. My point is that the unilateralist has a question to press.

(c) It is question-begging to dismiss unilateralism as incautious. If deterrence is delicate unilateral initiatives may well be reckless because they may well shift the delicate balance dangerously. But if deterrence is robust, as I think the unilateralist should argue, then it is nothing like so clear that unilateralism is reckless. On the contrary, the unilateralist can contend that the multilateralist is being incautious by resisting the only feasible way to initiate a disarmament process for the sake of a theory (the delicate balance theory) which is unsafe in its own terms, and arguably incoherent.

(C) A third objection to unilateralism comes from arms control. It is easy for unilateralists to claim that they welcome arms control agreements and desire to go further. But what if some or all of the important agreements have been and can be secured only by a posture of Western military strength incompatible with the deep unilateral cuts favoured by the unilateralist? In that case a stark choice has to be faced between disarmament and arms control, a choice made more uncomfortable if the probability of arms control agreements is relatively high and the probability of initiating disarmament is relatively low.

There is, I think, a real difficulty here for the unilateralist. But there are difficulties in every position, and three points on the unilateralist side seem to be in order. First, those arms control agreements of most obvious value to humanity, such as the Partial Test Ban Treaty and the Non-Proliferation Treaty, do not seem to presuppose any particular military strength on either side. Second, those arms control agreements which are intimately bound up with the military balance look far more important if one thinks deterrence is delicate than if one takes the view that it is robust. For example, the treaty limiting anti-ballistic missile defences is important in orthodox terms partly because unrestricted defences can be thought of as impairing the certainty of retaliation and so the reliability of deterrence. Such niceties loom less large if deterrence is robust. Third, arms control is in deep crisis, as the Director of the International Institute for Strategic Studies acknowledged before the Soviet invasion of Afghanistan.[7] Technology is digging a grave for arms control. The trends may be resistible, not least because of a natural alliance between those who favour arms control and those who desire to set some ceiling to military spending. But the future of arms control is far from secure, so the unilateralist can hardly be accused without qualification of sacrificing arms control for an uncertain prospect of disarmament.

(D) A fourth objection to unilateralism concerns blackmail. The issue is difficult to formulate but I take the gist of it to be this. Over and above the grand strategy of deterrence, the protecting of vital and dramatic interests such as West Berlin, there is a day-to-day grind of international politics in which resolve, perceived strength and perceived weakness count for something. If the West were to make deep cuts unilaterally, then we would be exposed to potentially damaging pressure even if deterrence were not threatened. It would be that much more difficult for the West to assert and concert the interests of freedom and of the nations and nation groups that make up the West.

If I understand the point aright, the real issue here is defeatist unilateralism. If the West made deep cuts not out of strength and conviction, as a bold initiative aiming to start a constructive disarmament process, but out of weakness and concession, as something forced upon governments by ignorant and fearful public opinion aiming to somehow remove (say) East Anglia or Scotland from the East-West confrontation, then unilateralism might well damage the West even if it did not touch deterrence. It would be damaging because it would be a sign of waning morale, of vanishing commitment to Western values and solidarity. In the Soviet Union it would strengthen the hand not of those who desire to eliminate needless East-West tension but of those encouraged to the vigorous world-wide promotion of Marxist ideology by every sign of demoralisation in the capitalist opponent.

If this is the issue it is a difficult one for the unilateralist. Fears for the lives of the people one loves have played a considerable part in the current revival of unilateralism. The mobilising of intensely local sentiment is important for a popular movement. And there may be unilateralists who are demoralised, or think a weakening of Western morale would be good for whatever they happen to cherish. Yet it would be a travesty to represent defeatism as the mainspring of unilateralism. If the unilateralist is to answer what lies (I take it) behind the worry about blackmail, then he needs to distinguish sharply between the reasonable fears we all have, which are a perfectly good reason for desiring the eventual abolition of nuclear weapons, and the argument for bold initiatives which is the real strength of his challenge to multilateralist orthodoxy.

(E) A fifth objection to the unilateralism for which I have been arguing is concerned with globalism. My argument pictures deterrence as a robust protector of the West's truly vital interests — the US heartland, Western Europe, Japan, oil supplies — but as powerless to determine the fortune of more marginal interests, most notably of anticommunist regimes, attacks upon which seem most unlikely to precipitate major war. To picture deterrence thus is, it might be argued, simplistic and/or irresponsible: simplistic because there are grey areas in which the extent and potency of Western commitment is far less certain, far more a function of our readiness to struggle than is the case with (say) West Berlin; irresponsible because our opponent has global ambitions which threaten the liberties of free peoples far beyond the borders of the indisputably Western bloc.

The grey-areas argument is essentially bound up with the belief that deterrence is delicate. If this belief is correct, if deterrence is a matter of achieving just the right amount of probability of reprisal which may escalate, then there are no obvious territorial limits to the operation of deterrence. The nuclear umbrella can in principle be extended across the whole free world. The problems of extending it thus far are essentially problems of technique and of political will. But if deterrence is robust, extending the deterrent beyond its natural limits is impossible. Short of entering into the kinds of bonding and integration that link the US, Western Europe and Japan, deterrence cannot extend to all those regimes one might like to support. If they are to be supported, it must be by means other than the robust deterrent deployed by NATO. If robust, deterrence eliminates grey areas: not, alas, in general, but for the purposes of deterrence.

Deep Cuts are Morally Imperative

As for the charge of irresponsibility, one need not be hostile to Western ideals about the liberty of free peoples to doubt the prudence of seeking one military strategy for the whole world. A series of American theories have aimed to negate the natural differences between America's most enduring and most adventitious allies. For example, the idea of limited war to which Henry Kissinger contributed in the late 1950s abstracts from the political to a military chess-board on which all foreign military commitments of the US are described in a uniform terminology. The problems of seeking to protect South Vietnam by military means and the problems of guaranteeing the security of Western Europe are pictured by limited war theorists as being of essentially the same kind. One is hardly being irresponsible if one doubts the realism of such conflation of unlike things. More recently, Dr Kissinger has been an influential exponent of the view that the West should seek to emphasise linkage between its various quarrels with the Soviet Union. This is the limited war idea all over again. Anyone who thinks deterrence is robust rather than delicate is bound to view the idea of military linkage with profound misgiving, for such linkage goes against the grain of deterrence's limitations.

I have laboured this issue, which may seem excessively academic and excessively American, because it contains a point of great importance for the nature and ambitions of unilateralism. The misgivings about military globalism that I have sketched are widely shared in Europe by multilateralists, and are voiced with considerable trenchancy by the allies in their dealings with the US. Unilateralists could perfectly well make common cause with multilateralists on this issue, and contend that unilateralism represents a very substantial programme for the concerted pressing of European realism upon our senior ally. This does not seem to be happening. Instead, the unilateralists seem to drift into a shadowy linkage theory of their own, denouncing the US for its conduct in Latin America and intimating that Europe should distance itself on all matters from its great ally. *A unilateralist need not say this!* To me it appears both bad tactics and a reason for profound disquiet that so many unilateralists appear to oppose themselves in deep principle to the Atlantic alliance.[8] It is bad tactics because the natural interest of Europeans in preserving the alliance is great and, I suspect, recognised as being great by most of the people who must be persuaded of the merits of unilateralism if political influence is to be gained. Far more serious, the animus against America, the readiness with which so many unilateralists are prepared to speak as though the US and USSR were powers of the same kind, makes one wonder what values inform a unilateralist that he can so easily forget the difference between totalitarian and liberal political order. In short, I believe that the unilateralist can make a strong reply to the charge that he is failing the West's global responsibilities; but one great obstacle to a whole-hearted trust of current unilateralism is its tendency to make instead of this reply cheap anti-American polemic.

(F) A sixth objection to my argument is that the contrast I have drawn between delicacy and robustness is too stark and schematic. Should we perhaps say that deterrence is robust *within limits*? Should we allow for the possibility that it is robust at one level (for example at the strategic nuclear level), but much more delicate at another (for example at the theatre level)?

There is certainly some truth in the view that there are limits to the robustness of deterrence. If one side had, say, two thousand nuclear weapons and the other had none, or only one, then it would be wildly unreasonable to regard the two sides as capable of mutual deterrence. But it would be a mistake to infer from this that the real question as between the delicate balance and robustness theories of deterrence is a highly technical question of exactly how much is enough to deter the opponent. If deterrence is robust, the margins of error that can be tolerated without dangerous upset are very wide indeed. I have been arguing that if deterrence works at all it does so by motivating the deterred to avoid certain dangers, *not* as the outcome of a highly technical computation of interests and probabilities *but* like the plague. If there is even a fairly low probability of war breaking out as a result of some action, then that action is to be avoided. If this is indeed how deterrence operates, then the margins of error are wide. There are limits to the robustness of deterrence but within those limits there is much latitude.

This conceptual point is important to my argument because it affects the balance of competence as between experts and laymen. If deterrence is delicate then it is extremely difficult for the ordinary citizen to make any responsible input into the decision-making process, since he lacks the information necessary for determining whether his input will be destabilising. But if deterrence is robust then he is not so inhibited, precisely because the margins of error are relatively generous. The reader will have noticed that I have been arguing for deep cuts without attempting to say which weapons systems should be cut. The reason is that I have been trying to intimate in this essay that there is a level of strategic debate to which a wide public can contribute, the level at which one discusses the fundamental assumptions of strategy. If I

Deep Cuts are Morally Imperative

am right, there remains inevitably a highly detailed and technical argument to be conducted about how the cuts should be made in such a way as to strengthen the hand of those in the Soviet Union who may favour reciprocal cuts. That argument will not take place without a major shift of public opinion. I have been arguing that the shift can be in a responsible direction. If the real question were, exactly how much is enough? then this wider public would be disqualified from responsible contribution to the debate.But this is not the real question and we, the wider public, are not disqualified.

As for the suggestion that deterrence may be robust at one level, delicate at another, this seems to presuppose that deterrence is delicate. My point has been that deterrence is indivisible, that a European adventure is ruled out by the brute and very stable fact that any war in Europe would be extremely hard to keep limited. I therefore think it most unlikely that deterrence is robust at some levels but not others.

(G) The last objection I shall discuss to my unilateralist argument is that I have said nothing about the consequences of failure. What if the West made deep cuts and the USSR failed to make any constructive response?

The possibility of failure cannot be excluded. If deep cuts were to fail, two sorts of consequence would in my opinon result. First, our bargaining position as defined in the current orthodoxy would be worse than it now is. We would have given away something for nothing, our enemies would have been strengthened by our actions, the prospects of arms control as we have known it being reinstated on terms congenial to the West would have been dimmed. Those who think deterrence delicate will find this a much more daunting turn of events than those who reckon deterrence is robust. Second, our opponent would have clarified matters grimly. At the moment many of us, not obviously irrational, have an open mind about the nature of the Soviet threat: we are able to argue that our opponent may be more of an inscrutable, ambivalent, intensely cautious power open to offers than a demonic force for evil. If a great Western initiative were made in good faith and were to fail, the voices interpreting Soviet behaviour in the harshest of terms would be far less easy to controvert. Perhaps all would not be lost. One might well argue that the attempt was worthwhile but the failure less meaningful than might be thought because it constituted such a break with what has always been normal in statecraft.[9] But, however convincingly one could moderate the lesson of a failure by explaining it as a noble but too novel experiment, great damage to East-West relations would have been done.

Deep Cuts are Morally Imperative

The risks have, I think, simply to be admitted. Every position on deterrence has its difficulties. The position for which I am arguing has in compensation the twofold advantage that it would be an honest attempt at withdrawal from a morally untenable stance, and that it is very far from being clear that the Soviet Union would fail to reciprocate.

To sum up the argument so far: I have claimed that deterrence is immoral because it is the conditional intention to wage all-out nuclear war. The laudable aim and well-intentioned complexity or strength of deterrence do not lead me to qualify this judgement, but it is not always right to desist at once from an action on recognition of its immorality. There are massive moral complexities in withdrawing from the morally impossible position into which we have entered, and more than one route to the abandonment of unrighteousness needs to be considered. But multilateralism seems to envisage permanent reliance upon deterrence, permanent persistence in iniquity, and this makes the investigation of unilateralism morally imperative.

The unilateralism I favour rests on an argument that has been current for a long time to the effect that deep cuts in current force levels should be tried in an attempt to initiate a process of disarmament. Anyone sympathetic to this argument has to confront the well-developed thinking that finds expression in Western deterrence. In this essay I have tried to meet head on a basic tenet of strategic orthodoxy, the delicate balance theory. I have tried to give reason for believing that deterrence is robust and have suggested that this enables the unilateralist to answer the multilateralist on the latter's strongest ground.

In conclusion I comment briefly on four aspects of the current debate which are important to my theme but less central to my argument. First, what about the cost of deterrence in a world so desperately in need of measures to feed the hungry? I make no apology for discussing nuclear weapons without reference to the starving. The probability of even large-scale nuclear disarmament freeing resources for the real needs of the needy is somewhat low. It is not clear that resources can, economically speaking, be transferred from the military to the non-military sector. Even if they can, there is no guarantee of their being devoted to the needy rather than to the politically influential classes of the disarming powers, always eager to reduce their tax burden and to increase their personal welfare. And even if the resources can somehow be got to those areas of the world in which the needy starve, political reform of these regions seems a vital and exceedingly difficult prerequisite of massive help to the truly needy as distinct from the understandably aspiring. I mention these difficulties not to deny that stark suffering is one of the great political issues of our time, but because the argument for unilateralism does not depend on the highly uncertain prospect of achieving redistribution of resources from defence to the needy. It depends on the urgent duty to withdraw from our iniquitous intention to wage wicked war. One blurs and softens the issue by bringing in the needy.

The second aspect concerns the nature of the deterrent. I have claimed that we in the West rely upon graduated deterrence in order to delay the use of nuclear weapons in any battle. Some would say that this formulation is out of date. They argue that the US has been moving towards a war-fighting capability requiring the early use of nuclear weapons since 1974, when Secretary of State Schlesinger announced a shift in targeting policy, and more especially since President Carter's Presidential Directive 59 placed heavy emphasis on nuclear strikes against military targets in the USSR. These critics often conclude that the US is abandoning deterrence in favour of war-fighting, and argue from the distinction between strategic and theatre missions that the US is preparing to fight a nuclear war in Europe which leaves the US unscathed.

Such a view is mistaken for at least four reasons. First, everyone knows that the prospect of keeping nuclear war limited is poor. The Americans would be mad if they were intending to fight a nuclear war in Europe that left the US unscathed. Even the most hawkish are not mad. Their aim is not to fight a limited nuclear war but to minimise the likelihood of war and to strengthen intra-war deterrence. They judge (I think wrongly) that the best means to this end is to develop what critics call a war-fighting capability. Second, deterrence cannot be equated with the West's traditional flexible-response posture. The USSR has a radically different kind of deterrent: Russia's readiness to use nuclear weapons large and small early does not prevent this being, in aim and effect, a deterrent. If the US moves to something more like the Soviet system, it is altering its means of deterrence not abandoning deterrence. Third, the introduction of systems at some points on the escalation ladder that some of us would judge inimical to deterrence does not abolish so much as weaken and endanger the West's graduated deterrent. Shifts of emphasis in doctrine and mistakes in weapons deployment do not necessarily destroy flexible response as we have known it. Fourth, it is absurd to suppose the US administration is a monolith, acting and pronouncing with one mind and one voice. Many military personnel have never felt happy with Western ideas of deterrence. Like their Soviet colleagues, they have felt that there is no substitute for sheer strength. They constitute a formidable lobby. It would be a mistake to equate political concessions to their gut feelings with the well-considered strategy of the Western alliance.

The third aspect concerns whether the unilateralist should support higher conventional force levels, and the higher defence budgets that might be required to achieve these higher force levels? Since the 1950s NATO has had its defence on the cheap because early resort to nuclear weapons is cheaper economically and politically than attempting to reach a reasonable East-West force balance at the conventional level. Should one who objects to deterrence on moral grounds support, even actively campaign for, increased conventional forces? I lack the space to discuss this issue at the length it deserves, confining myself to two over-brief remarks. First, the unilateralist needs to be aware that NATO thinks is conventional force levels on the low side even relative to its existing defence posture, even with the present plan that could well involve the West being first users of nuclear weapons in a European war. The current NATO demand for increased conventional forces derives from the strategic orthodoxy, from the belief in the delicate balance. One who views deterrence differently would be unwise to endorse uncritically NATO's desire for higher conventional forces. But, second, the unilateralist could well attempt to obtain an estimate of the force levels that NATO would require to give a no-first-use undertaking. If there is some attainable level of conventional forces which would enable the West to foreswear the first use of nuclear weapons in Europe then the contribution to nuclear disarmament might be so great as to be well worth supporting.

Finally, why do the unilateralists not talk about strategy? One important factor, I suspect, is that they do not believe that the USSR poses a serious military threat to the West and/or they view strategy as a rationalisation of technology and of military secrets rather than as an attempt to respond rationally to the Soviet threat. Such an attitude to strategy is understandable. The USSR would, for example, be crazy to think that its interests would be served by its coming to dominate Western Europe. It has enough difficulty controlling the relatively sluggish states of Eastern Europe, and one of its most fundamental post-war objectives — to prevent the reunification of Germany — would be frustrated if Germany were reunified even within the Soviet sphere of influence. A parallel argument concerning West Berlin is not plausible, but if Berlin were the only obstacle to removing the military danger in Europe one might well contemplate the evacuation of the city. And careful historical enquiry can suggest rather strongly that some at least of Western strategic writing is a rationalisation rather than the reason for certain military measures.

But the military dimension of affairs has a certain life of its own. If the USSR were faced with opportunities in Europe by ill-judged Western moves, it could perfectly well succumb to the temptation, not least for strategic reasons having little to do with what one would think of as real, non-strategic Soviet interests. Furthermore, the idea that deterrence is delicate has a compulsiveness and plausibility altogether deeper than any rationalisations of the moment. In one form or another it has been present ever since the West began to think about relying on nuclear weapons for a significant part of its safety in the late 1940s. It is more enduring and deeper than mere rationalisation. It requires an answer rather than a brush-off.

Above all, if unilateralism is to gain a serious place in the counsels of the West it must gain widespread agreement among thoughtful voters. It must go through the fire of critical debate, in which a wide variety of perceptions of the Soviet threat and of attitudes to strategy are brought to bear. It must speak to something more enduring and more infused with rationality than gut reluctance to have nuclear weapons in one's backyard and unexamined feeling that there is some truth of some kind in the plurality of mutually incompatible criticisms of NATO that are on offer. It seems to me inconceivable that unilateralism can survive this test without entering into serious strategic debate.

Notes

1. What of the saying 'Better red than dead?' Like other contributors to this book, I think we can avoid the choice, and in this essay I try to spell out what seems to me the most promising way to avoid both nuclear death and subjection to the rule of a totalitarian power. But what if one were faced with the stark choice red or dead? The Gospel makes no promise of our being spared such choices. It speaks rather of the Cross. So far as I can see, this must mean that we are required to undergo gross injustice that will break many souls sooner than ourselves be the authors of mass murder.

2. A good example is Harold Brown's speech to the US Naval War College, 20 August 1980, *The Objective of US Strategic Forces* (International Communication Agency, US Embassy, London, 1980).

3. There is not space in this essay to discuss British neutralism, that is the view that the UK should leave NATO and denounce the North Atlantic Treaty in order to distance itself from deterrence. This is because I would regard such a move as a disappointing last resort: better that the UK should remain within the alliance arguing for the unilateralism commended in this essay. Our getting out leaves the

rest of the world in the iniquitous position I have tried to describe. This essay is also silent on the decision to acquire a successor to the Polaris British independent nuclear deterrent, to which one might object on a wide range of counts including the likely impact on our contribution to NATO conventional forces and the dubious value of a third centre of Western decision-making in addition to Washington and Paris.

4. Elizabeth Young points out that there is another option, ignored in my essay, of multilateral disarmament negotiations. For my limited purpose, I set this aside because the achievement of disarmament agreements is bound to be even harder than the attainment of arms control agreements.

5. I do not wish to imply that all arms controllers aim to secure the moral objective that interests me. Many of them would dismiss it as utopian. But this cannot be proclaimed too loudly, for governments point to arms control when challenged about the sincerity of the declaration of all the Permanent Members of the UN Security Council in favour of the eventual abolition of nuclear weapons.

6. See 'East-West Dissidents – a Conversation' (between Thompson and Roy Medvedev), *END Bulletin no. 1* (Bertrand Russell Peace Foundation, Nottingham, 1980), pp. 4-6.

7. Christoph Bertram 'Arms Control and Technological Change: Elements of a New Approach', in *The Future of Arms Control*, Part II, Adelphi Paper, no. 146 (International Institute for Strategic Studies, London, 1978).

8. Remember that the North Atlantic Treaty is *not* a treaty about nuclear weapons.

9. I am grateful to Michael Donelan for this point.

6 THE CASE FOR NEGOTIATED DISARMAMENT
Roy Dean

Background

In the years following the resolution of the Cuban missile crisis in 1962, there were hopeful signs that an era of negotiation had begun which would enable East and West to build a structure of peace. The Partial Test Ban Treaty of 1963 was the first concrete indication of what might be achieved. There followed the signing of the Nuclear Non-Proliferation Treaty in 1968. The Strategic Arms Limitation Talks (SALT) opened in 1969 and yielded two important agreements in 1972: the Anti-Ballistic Missile (ABM) Treaty and the Interim Agreement on the Limitation of Strategic Offensive Arms. The US Senate ratified both treaties quickly and negotiations began in the same year for a follow-on to the Interim Agreement.

Multilateral negotiations for the Biological Weapons Convention proposed by Britain were successfully completed in 1972. The Mutual and Balanced Force Reduction (MBFR) talks in Europe opened a year later. In 1974 the Vladivostok Accord seemed to pave the way for SALT II. Confidence-building measures (CBMs) providing for prior notification of military manoeuvres and movements in Europe were agreed at the 1975 Helsinki Conference on Security and Co-operation in Europe (CSCE), which also marked the culmination of steady improvements in East-West relations over the previous few years. The most dramatic symbol of these improvements was the Apollo-Soyuz link-up in the same year.

The hopes to which this series of events gave rise have been dashed by developments in the second half of the 1970s. The difficulties began with the Soviet intervention by proxy in Angola; subsequent interventions in Ethiopia and, most recently and blatantly, in Afghanistan have progressively increased world scepticism about the peaceful intentions of the Soviet Union. The sustained increase in all aspects of Soviet military power through this period has led to Western doubts about the effectiveness of arms control as a contribution to international security. The whole arms control process has come under scrutiny on both sides of the Atlantic. The basis of trust for arms control agreements has been seriously eroded.

Nevertheless the majority of the world's leaders remain convinced that negotiated arms control and disarmament is the only way to curb the arms race, in both nuclear and conventional weapons, and to create greater national and international security at a lower level of risk and expense. Presidents Brezhnev and Reagan have both made statements expressing their desire for strategic arms reductions. West European leaders have placed particular emphasis on the limitation of theatre nuclear forces on both sides in the framework of the SALT negotiations. The non-aligned states continue to attach the highest importance at the United Nations to nuclear disarmament by the superpowers.

Ethical Considerations

Barrie Paskins has discussed in Chapter 5 the ethics of nuclear deterrence. The first question to examine here is the part played by ethical considerations in the arms control process. There is no doubt that they have been influential in the prohibition of weapons deemed to be particularly abhorrent — such things as biological weapons and 'inhumane' conventional weapons. But one fact emerges clearly from the history of the disarmament negotiations: they are primarily concerned with security.

It could be argued that the government of a country has a moral duty to protect its people from attack; it would be an immoral and irresponsible act if, by failing to deter a nuclear attack, the government allowed the country to be obliterated. So where the negotiators have to make a policy judgement involving an element of calculated risk — and this applies with particular force to nuclear weapons because of their appalling destructive power — morality and security should go hand in hand. But equally, moral imperatives demand that the level of armaments should be reduced in order to release resources for urgent economic and social needs.

It would be unrealistic to expect the major military powers to base their negotiating approach on ethical considerations. The best that can be hoped for is that both sides will go into negotiations with a sense of responsibility, negotiate in good faith, and comply with their treaty obligations. The force of world opinion is a particularly strong factor for any state contemplating a breach of an arms control agreement. There is another ethical constraint inasmuch as the negotiating parties must have credibility; governments must ensure that their defence programmes are not incompatible with their arms control posture.

The Case for Negotiated Disarmament

The leading members of the two military blocs have their share of 'hawks' and 'doves', not only in government but also in non-governmental circles. The hawks see arms control as a diversion from the serious business of military preparations; the doves call for instant disarmament and think arms control is a cosmetic exercise to agree on permissive restraints with no real reductions.

In the middle of the debate stand those who might be termed the 'owls' — the arms controllers who argue that the world is overarmed, that security can be assured at a reduced level of forces, and that there are enough common interests to bring to the negotiating table even those countries which are in a state of military confrontation. The clash between the conflicting interests sometimes produces an ambivalent government policy — for example in the Soviet Union — in which precept may differ noticeably from practice. There is a different complication in democratic states where a change of government can lead to the introduction of a completely new arms control policy as in the US.

The battle rages most fiercely in the White House and the Kremlin. The outcome usually depends on each side's perception of how far it can trust the other; aggressive behaviour by one side inevitably reinforces the hand of the military on the other side, thus having a ratchet effect on the arms race. Attitudes to arms control are also important. Whereas openness builds confidence, secrecy breeds distrust.

Deterrence and Disarmament

In Chapter 4 Arthur Hockaday has explained the theory of deterrence: the policy of persuading the Soviet leadership that the risks of launching aggression against the West outweigh any possible gains they could hope to achieve. The context of deterrence is also well illustrated in an essay in the 1981 Defence White Paper[1] on 'Nuclear Weapons and Preventing War' included as an appendix to Chapter 4. In a world in which both East and West possess nuclear weapons, peace depends on a 'balance of terror' being maintained.

But that cannot be the end of the story, otherwise both sides would be condemning themselves to an endless arms race. This has to be prevented by negotiated arms control and disarmament — first to make the 'delicate balance' more stable, and then to bring down the appallingly high level of armaments on both sides. In Britain, successive governments have seen deterrence and disarmament as both necessary and complementary ways of achieving the overall objective of peace and

120 *The Case for Negotiated Disarmament*

security.

The problem of nuclear weapons and deterrence has caused differences of view within the Christian community in Britain, not on the need to control and eventually eliminate both nuclear and conventional forces, but on the practical means of bringing this about. This question came up in a resolution on unilateral nuclear disarmament by Britain discussed by the Assembly of the British Council of Churches in November 1980. In his address the Archbishop of Canterbury, Dr Robert Runcie, said:

> I respect the sincerity of those who take this line but I must say that I doubt the exemplary power of this gesture which would be made as we still expect for the foreseeable future to be sheltering under the American nuclear umbrella, and I fear it might serve to destabilize a balance which has undoubtedly contributed to the peace of Europe for thirty-five years.[2]

This is where the advocates of multilateral disarmament — who support deterrence and 'negotiating for peace' — part company with the so-called 'peace movement', which is heavily weighted in favour of unilateral disarmament. They see an immediate danger that talk of unilateral moves by Britain will encourage the Russians to block any negotiations — as they did in 1979 when NATO proposed talks on limiting long-range theatre nuclear forces — in the belief that if they wait long enough the West will disarm on its own, without seeking reciprocal reductions by the East. Any one-sided reduction by the West, they say, would weaken its ability to deter aggression, and could therefore make war more likely rather than prevent it. Nor would Britain be safer without nuclear weapons; Soviet nuclear missiles have been targeted on the UK for over twenty years, and their military planners are hardly likely to change their policy, given the key industrial and strategic importance of the UK.

Some analysts have concluded that the logical consequence of unilateral nuclear disarmament would be neutrality. On the question of British renunciation of nuclear weapons and withdrawal from NATO, an academic expert has written:

> A unilateral departure by Britain would heighten by perhaps a factor of ten the near-term danger of European war, especially if she renounced the Rome Treaty at about the same time. Furthermore, it would accentuate anxieties in those countries (from Sweden to Israel

The Case for Negotiated Disarmament

and far beyond) which, though outside the two nuclear blocs, relate their own security to the balance between them. The implications for the proliferation of independent deterrents could therefore be adverse.[3]

Even if a nuclear-weapon-free zone were created 'from the Atlantic to the Urals', as proposed by European Nuclear Disarmament,[4] the effect would be one-sided because the accurate and mobile SS.20 missiles would continue to pose a threat to the countries of Western Europe from sites on Soviet territory east of the Urals. Moreover, the idea of a Europe free from nuclear weapons takes no account of conventional forces where the existing imbalance, without the restraint imposed by nuclear weapons, would be a source of great instability. A Europe without nuclear weapons would be an infinitely more dangerous place, unless conventional disarmament can also be achieved.

Peace Movement

There is no evidence to suggest that unilateral disarmament by Britain would greatly reduce the number of nuclear weapons in Europe or help to prevent further proliferation elsewhere in the world. The Russians have made it clear that they are opposed to unilateral disarmament for themselves. In a recent analytical article two policy-makers wrote:

> The Soviet Union cannot undertake the unilateral destruction of its nuclear weapons (and indeed has no right to do so, as it is responsible to the peoples of the whole world for peace and progress). To do so would mean disarming in the face of the forces of war and reaction.[5]

Nor is the Soviet Union concerned with ethical objections to the use of nuclear weapons. In the same article the authors stated: 'Whilst speaking against the use of nuclear weapons, the Soviet Union does not exclude the possibility of using them in extreme circumstances . . . Marxist-Leninists decisively reject the assertions of certain bourgeois theoreticians who consider nuclear missile war unjust from any angle.'

The idea that the 'peace movement' can be extended to the Soviet Union is wishful thinking. In the early 1960s, CND protesters who unfurled their banners in Moscow were speedily rounded up by the security police. The courageous group of people trying to monitor the implementation of the Helsinki Final Act have been ruthlessly suppressed. It is doubtful whether the Kremlin would allow demonstrations for Soviet disarmament in Red Square, or permit news of arms control proposals by Western states to reach the Russian people, although it encourages general petitions in favour of peace and disarmament. The apparatus of the state ensures that millions of signatures can be obtained for such petitions.

In his dialogue with Professor E.P. Thompson on the prospects for a European nuclear-weapon-free zone the Soviet historian, Roy Medvedev, pointed out that no kind of 'popular initiative' or mass movement against nuclear weapons can be mobilised in Eastern Europe. 'We have,' he wrote, 'no public movements independent of the control of the Party and the Government.' The same point was made in a discussion[7] between Professor Thompson and a Czech dissident writing under the pseudonym of 'Vaclav Racek'. Without freedom of information and expression, the public cannot form a judgement and lobby their government to adopt a particular policy.

Past Experience

Ramsay MacDonald was proved right when he said in 1934, in the context of the World Disarmament Conference, that 'disarmament by example simply does not work'. The events of the 1930s showed only too clearly that unilateral restraint led to disaster. The experience since 1945[8] also shows that unilateral disarmament is not a realistic proposition, whether or not an effective military balance already exists. Such acts of self-denial as have been taken from time to time, for whatever reasons, do not appear to have produced satisfactory responses.

For example, in June 1946 the United States voluntarily offered to give up its monopoly possession of nuclear weapons in a bid to abolish them completely. The Baruch Plan envisaged their destruction and all atomic plant and fissionable materials being brought under the control of an international authority with no veto on its powers. The Soviet Union wanted the nuclear stockpile to be destroyed before the authority was set up, and insisted on the right of veto because it was then in very much of a minority position in the Security Council. This created a stalemate which persisted until the Soviet detonation of an atomic bomb in September 1949 put paid to the unique opportunity of stifling the nuclear arms race at birth.

After the Second World War the United Kingdom destroyed its

offensive chemical weapons capability; the Soviet Union continued to build up its own and to develop a chemical weapons war-fighting doctrine. In July 1956 the Soviet Union accepted a United States proposal that each should reduce its forces to 2½ million as a step towards greater disarmament, but no agreement was sealed. According to the International Institute for Strategic Studies,[9] the total US armed forces had fallen to 2,022,000 by 1980, while the level of Soviet armed forces stood at 3,658,000 (not counting some 500,000 internal security, railway and construction troops). In 1960 Britain abolished conscription service, followed by the United States after the Vietnam War; the Soviet Union continues to maintain a large conscript army and massive reserves.

In the early 1960s the United States decided not to proceed with anti-satellite weapons; the Soviet Union began experiments in 1968 which paved the way for an arms race in space. Between 1968 and 1974 the United States carried out a planned reduction of its defence budget; the Soviet Union accelerated and in 1971 overtook the US as the largest military spender.[10] The withdrawal of British forces east of Suez was completed in the 1970s, leaving a vacuum which the Soviet Union was only too happy to fill. In 1977 President Carter adopted a policy of voluntary restraint on American conventional arms transfers, virtually cutting off the supply of US weapons to all but a few Third World countries. Once again the Soviet Union moved in and, according to the Stockholm International Peace Research Institute (SIPRI),[11] quickly became the largest supplier of arms to the poorest countries of the world in Asia, Africa and Central America.

Since 1968, when the superpowers agreed to start negotiations on nuclear disarmament, NATO had deliberately refrained from deploying new theatre nuclear forces in Europe. However, in the mid-1970s the Soviet Union started to introduce the powerful new SS.20 missile and the Backfire bomber. The US unilaterally withdrew 1,000 nuclear warheads from Europe as part of the TNF modernisation/arms control decision of December 1979, with no response from the Eastern side. While NATO's collective military expenditure fell in real terms by 9.4 per cent between 1969 and 1978, that of the Warsaw Pact rose by 31.5 per cent.[12]

Disarmament by Negotiation

In Chapter 3 Bruce Kent has advocated 'leading by example'. But this ignores the political realities of the international situation in which the disarmament negotiators are operating. The weight of experience suggests that unilateral action by country A removes the incentive for country B to negotiate. The argument that multilateral negotiations are only a series of unilateral moves is at odds with the real world of arms competition. Multilateral disarmament is more likely to succeed when two countries perceive a mutual interest in giving up a degree of armament, each being assured that its potential adversary is making similar concessions. In this way there are no hostages to fortune of the kind that unilateral disarmament produces. Both sides move together and undiminished security is maintained.

The late Earl Mountbatten set out the rationale for negotiating nuclear disarmament in an address in Strasbourg on 11 May 1979, when he warned against the dangers of an arms race leading to nuclear war. He said:

> But how do we set about achieving practical measures of nuclear arms control and disarmament? To begin with, we are most likely to preserve peace if there is a military balance of strength between East and West. The real need is for both sides to replace the attempts to maintain a balance through ever-increasing and ever more costly nuclear armaments by a balance based on mutual restraint. Better still, by reduction of nuclear armaments, I believe it should be possible to achieve greater security at a lower level of military confrontation.[13]

These conclusions were borne out in the UN Secretary-General's study on all aspects of nuclear weapons which was presented in 1980. On the question of disarmament the report stated:

> Nuclear disarmament, if it is to be comprehensive and meaningful, will have to be pursued in a global context. It is to be understood that in the first stage, the two major nuclear weapon states have to make the initial reductions in their nuclear arsenals and to effect substantive restraint on the qualitative development of nuclear weapon systems. They should seek to achieve this objective in the framework of bilateral negotiations.[14]

The Machinery

The United Nations Charter conferred specific responsibilities in respect

The Case for Negotiated Disarmament

of arms control and disarmament on the Security Council and the UN General Assembly. The machinery for serious negotiations has been refined over the last thirty-six years.[15] A constant element has been the annual review by the United Nations General Assembly First Committee of discussions and negotiations taking place in smaller bodies specifically established to consider the many political and technical problems which make up the question of disarmament.

A noteworthy landmark in the history of the United Nations General Assembly was its first Special Session on Disarmament in 1978 (UNSSD I). This was a non-aligned initiative designed to involve all countries in the disarmament debate and to work out an acceptable disarmament strategy. The Final Document[16] which the member states adopted by consensus was the most comprehensive statement on disarmament ever accepted by the world community. Its conclusions remain valid and should be the starting-point for the second Special Session in 1982 (UNSSD II).

UNSSD I endorsed arrangements for an expanded 40-member Committee on Disarmament (CD), with the participation of all five nuclear-weapon states, as the 'single multilateral disarmament negotiating forum of limited size taking decisions on the basis of consensus'. The CD's enlarged membership ensures better representation of different regions and viewpoints. The chairmanship is rotated on a monthly basis. At the conclusion of its summer session the CD submits a report to the General Assembly, which forms the background to the disarmament debate in the First Committee.

The pattern for international action on disarmament was prescribed in paragraph 8 of the Final Document of UNSSD I, which stated:

> While the final objective of the efforts of all states should continue to be general and complete disarmament under effective international control, the immediate goal is that of the elimination of the dangers of nuclear war and the implementation of the measures to halt and reverse the arms race and clear the path towards lasting peace.

To meet this historic challenge is clearly in the political and economic interests of all nations; it is also essential in order to ensure genuine security and a peaceful future for all peoples of the world. But as the Final Document recognised, the achievement of these objectives is not a simple matter. Progress towards world disarmament would depend on two related factors – increasing international confidence

126 *The Case for Negotiated Disarmament*

and trust, and successfully negotiating a number of specific arms control measures.

The first step is agreement on a planned and co-ordinated programme of action, moving by phases in both nuclear and conventional disarmament. This was the approach proposed by the United Kingdom and other Western states in a submission[17] to the Preparatory Committee for UNSSD I. It found expression in the Final Document as 'the elaboration of a comprehensive programme of disarmament encompassing all measures thought to be desirable'. The outline of such a programme[18] was drawn up by the UN Disarmament Commission in 1979, and the Committee on Disarmament was given the task of fleshing out the programme for discussion at UNSSD II.

Declarations Not Enough

The comprehensive programme of disarmament will call for practical and businesslike negotiations, not declarations of intent. The 1970s were declared the 'Disarmament Decade', yet world military expenditure increased faster than ever in this period, with the developing countries increasing more in proportion than any other group. In 1971 the Indian Ocean was declared a 'Zone of Peace', but this did not prevent the Soviet Union from establishing bases at Aden and Berbera, stepping up its naval presence in the area, and invading one of the hinterland states. The Declaration of Ayacucho by eight Andean states in 1974 on limiting armaments in their region has not yet been translated into effective action.

Every year the Soviet Union and its allies issue resounding 'declarations' on peace and disarmament which have no meaning in real terms, and follow up with declaratory proposals in the UNGA First Committee which divert attention from the need for serious negotiations. These declarations in the 1970s were profoundly cynical because they masked a massive Soviet military build-up; they were also unhelpful because they lulled people into believing that some action on disarmament was actually taking place.

The growing tendency for 'disarmament by declaration' is one of the side-effects of a situation in which most governments do not have the will or the intent to reduce their own forces, but feel better if they have put their name on a statement urging everybody to disarm. The Final Document of UNSSD I itself contains recommendations to which many members of the United Nations have paid scant attention. It is

The Case for Negotiated Disarmament

therefore important to recognise the limitations of what can be accomplished through UN discussions. Neither the General Assembly nor the Disarmament Commission is a forum for negotiating multilateral arms control agreements; only the Committee on Disarmanent can do this. On the other hand, the advantage of the UN system is that it involves all countries in the disarmament process; not just the nuclear-weapon states but also the developing countries which are the major customers for arms and the instigators of more than 130 wars fought with conventional weapons since 1945.

Strategic Arms Limitation

The United Nations has always given the highest priority to nuclear disarmanent, placing the major responsibility on the superpowers. The earliest proposals to halt the growth of stockpiles of strategic weapons were part of wider plans for general and complete disarmament (GCD) put forward by the United States and the Soviet Union in the 1950s. The proposal to consider the question of strategic arms limitation as a separate issue was made by the United States in January 1964. When the Non-Proliferation Treaty was opened for signature on 1 July 1968 the US and USSR jointly announced that they had reached agreement to begin bilateral discussion. These SALT discussions started in late 1969.

The Anti-Ballistic Missile Treaty — which by restricting ABM systems to one target on each side is a key element in ensuring effective deterrence — and the Interim Agreement were signed in May 1972 as the conclusion of SALT I. The SALT II negotiations began six months later and were charted in the Vladivostok Accord of 1974. President Carter made this his first foreign-policy objective when he assumed office. However, a proposal by the United States in March 1977 for 'deep cuts — up to 50 per cent' in strategic arms was rejected by the Soviet Union. Finally the SALT II agreement was signed by Presidents Carter and Brezhnev in Vienna in June 1979.

The SALT II agreement was seen as a carefully balanced compromise between the differing interests and attitudes of the two sides. It broke new ground in placing equal ceilings on the strategic systems of both sides and in applying qualitative limits. Its central feature is an effective limit of 2,250 on the total number of strategic delivery vehicles both sides may possess, with sub-ceilings on different elements within the aggregate. It also includes limitations on new types of inter-continental ballistic missiles (ICBMs) and on the modification of existing types. It provides for the dismantling or destruction of arms surplus to the limits set by the treaty.

The US Administration worked patiently to build up the two-thirds Senate majority required for ratification of the agreement. Sadly, the Soviet invasion of Afghanistan in December 1979 destroyed any hopes of Senate support, and President Carter judged it necessary to defer further action on ratification. On assuming office President Reagan called for a review of US arms control policy across the board, centred on SALT. He has since said that he is prepared to negotiate as long as necessary to reduce the numbers of nuclear wepons to a point where neither side threatens the survival of the other. President Brezhnev has stressed the importance of continuing the SALT negotiations in order to achieve genuine nuclear disarmament. Observers have drawn some encouragement from statements by US spokesmen showing renewed interest in reductions rather than limitations.

It would be no exaggeration to say that the prospects for progress in the whole field of arms control and disarmament rest on the fate of SALT. A continuation of the SALT process is necessary not only because of the need to limit strategic arms, but also for the impetus this would give to other arms control negotiations, particularly those aimed at reaching a comprehensive ban on nuclear tests.

Theatre Nuclear Forces

The SALT negotiations were facilitated to some extent because the two superpowers had achieved broad strategic parity, at such a high level that equal reductions could be made without affecting their ability to destroy each other. This state of affairs does not, however, exist in the field of long-range theatre nuclear forces (TNF) of East and West covering the European continent. Here the Soviet modernisation begun in the mid-1970s with the introduction of the SS.20 missile and the Backfire bomber, led to growing concern in Western Europe that the comparable land-based forces of NATO — namely the American F111 and British Vulcan aircraft based in the UK — were ageing and becoming increasingly vulnerable to the new Soviet weapons.

The West Europeans took the view that they could no longer ignore the growing gap in TNF which had created a 4 : 1 advantage in the Soviet Union's favour. To maintain credible deterrence they needed to show the Soviet leadership that any idea it might have of fighting a

The Case for Negotiated Disarmament

'limited nuclear war' in Europe, and escaping unscathed, was quite untenable. Hence the decision in December 1979 to modernise long-range TNF with US ground-launched cruise and Pershing II missiles. At the same time the West also needed to create an equitable basis for negotiations with the Soviet Union on mutual limitations; without modernisation by NATO there would be no incentive for the Russians to negotiate.

With full European support,the US made an offer to the USSR in December 1979 to negotiate equal and verifiable reductions on the theatre nuclear missiles on both sides. The USSR rejected this offer on several occasions until July 1980, when it reversed its decision and agreed to talks. The first round of the bilateral discussions took place before the US presidential election in November 1980. The new US Administration reaffirmed its determination to pursue these negotiatins and announced plans to resume them in November 1981. There are two years in which to achieve results, since the NATO deployments are not due to begin until the end of 1983. NATO has made it quite clear that success in arms control could lead to cancellation of its cruise missile programme. But the task will be difficult because the sites set up for SS.20s already give the Soviet Union more missile warheads than are envisaged in the whole of the NATO programme. The objective will be the 'zero option' – no such missiles on either side.

Stopping Nuclear Tests

Although nuclear weapons have not been used since 1945, SIPRI has estimated that more than 1,250 nuclear weapon tests have been carried out in that period.[19] In the 1950s the hazardous radioactive fall-out from atmospheric tests caused mounting concern throughout the world. The Indian Prime Minister, Pandit Nehru, proposed a ban on testing in 1954. The three states then possessing nuclear weapons – the US, USSR and UK – responded by starting negotiations on a test ban in 1958. They voluntarily undertook a moratorium on testing when the talks began, but this ended when the Soviet Union suddenly resumed testing in 1961, causing the conference to break up without agreement.

Following renewed international pressure, the three Foreign Ministers met in Moscow in August 1963 and signed the Partial Test Ban Treaty banning nuclear tests in the atmosphere, in outer space and under water. The treaty could not be made comprehensive because the Soviet Union refused to accept the degree of international inspection which would have been necessary to detect and identify underground tests. The PTBT entered into force in October 1963 and now has over a hundred parties. Often dismissed as a 'clean air act', it has in fact played a very useful role in limiting the ability of states to carry out a testing programme.

After some years of inconclusive discussion in the Disarmament Committee in Geneva, the US, USSR and UK began talks in July 1977 on a comprehensive test ban treaty. The talks went well and the three negotiating parties made good progress in pursuit of an effective agreement. But they could not agree on duration and on technically sound means of ensuring compliance with a treaty. The negotiations were temporarily suspended in November 1980 until the new US administration could clarify its arms control policy.

There is no doubt that a test ban would curb the development of new nuclear warheads. Some experts believe that, by creating uncertainty about the reliability of warheads in existing stockpiles, it would also downgrade the importance of nuclear weapons and provide an incentive for their destruction. And by ending the privilege of testing which the nuclear-weapon states at present enjoy it would remove an element of discrimination in the Non-Proliferation Treaty. Finally, the adherence of near-nuclear states, such as India and Pakistan, would provide a helpful boost to the non-proliferation regime.

Non-Proliferation

Preventing the spread of nuclear weapons to further countries is essentially a political problem which must be tackled by diplomacy and international co-operation. Technical obstacles may delay proliferation but they are unlikely to stop a determined government from pursuing the nuclear weapons option, if it feels its security is gravely threatened. Any action by the UK on its own nuclear weapons is irrelevant to that decision, because the UK does not represent a threat to any non-nuclear-weapon state.

The risks inherent in nuclear proliferation were recognised by the UN in 1959 when the General Assembly, on the initiative of Ireland, took up the idea of an international treaty to avert the danger of an increase in the number of nuclear-weapon states. The increasing availability of nuclear weapon technology was underlined when France carried out a nuclear test in 1960, followed by China four years later. Serious discussions began in the Eighteen-Nation Disarmament Committee in 1964. The Non-Proliferation Treaty which resulted after four years of solid negotiation was endorsed by the UN General Assembly and opened for signature in July 1968.

The NPT represents a bargain between its nuclear-weapon state (NWS) depositaries and its non-nuclear-weapon state (NNWS) parties. The NWS undertook to negotiate towards their own nuclear disarmament. In return, and with the guarantee that their access to the peaceful uses of nuclear energy would not be restricted, the NNWS renounced the acquisition of nuclear weapons and accepted full-scope safeguards administered by the International Atomic Energy Agency (IAEA) to verify this renunciation.

The treaty entered into force in March 1970 and has attracted very wide support from the international community. To date, 115 of the world's independent states are parties, including many industrialised states with an unquestioned ability to develop nuclear weapons. No NWS party to the NPT has been in breach of its treaty obligations by helping to spread nuclear weapon technology, and no NNWS party has emerged as a NWS since the treaty came into force. The nuclear arms control negotiations mentioned earier (SALT, TNF, CTB) are part of the commitment by the NWS under Article VI.

Nevertheless the treaty has failed to win acceptance from two of the overt NWSs, China and France, and from a number of significant NNWSs with advanced civil nuclear programmes, who prefer to keep the nuclear weapons option open. The relatively few nuclear facilities which are not covered by IAEA safeguards agreements are in India, Israel and South Africa, none of which is a party to the NPT. Despite the allegations by Israel in justification of its raid on the Iraqi reactor in June 1981, the IAEA inspectors had found no evidence that Iraq (a party to the NPT) had plans to divert fissile material from its research reactor for weapons purposes. There is, however, a particular danger which stems from the loophole in the NPT whereby an NNWS party to the treaty might unwittingly help an NNWS non-party to acquire the materials and technology for a weapons programme. It therefore seems more likely that any proliferation which takes place will be by non-parties to the treaty (for example the Indian 'peaceful nuclear explosion' in May 1974). Present concern about the 'Islamic bomb' is based on the fact that Pakistan, a non-party to the NPT, is developing unsafeguarded facilities for uranium enrichment despite repeated assurances that its nuclear programme is for peaceful purposes only.

Widespread discontent at the slow pace of nuclear disarmament by the superpowers was expressed by NPT parties at the Review Conference held in 1980. There was also criticism of the Nuclear Suppliers Group for their restrictions on the transfer of specially sensitive technology, which some states held to be incompatible with the treaty's provisions on facilitating peaceful uses. It was not possible to agree on a Final Document at the conference; but there was no withdrawal by disgruntled parties from the treaty. It may be an imperfect instrument, but it remains the best available political basis for diplomatic efforts to halt the spread of nuclear weapons. Much will now depend on the success of the IAEA Committee on Assurance of Supply in meeting the needs of civil nuclear industries within the framework of non-proliferation provisions which are internationally acceptable.

The setting up of a nuclear-weapon-free zone (NWFZ) can contribute to non-proliferation and regional stability in regions where nuclear weapons are not already deployed. The concept has already been applied in agreements to prohibit nuclear arms on the sea-bed (the Sea-Bed Treaty of 1971), in outer space (the Outer Space Treaty of 1967), in Antarctica (the Antarctic Treaty of 1959) and in Latin America (the Treaty of Tlatelolco of 1967). The UK was the first nuclear-weapon state to ratify the Additional Protocols to Tlatelolco, thereby agreeing to respect the status of the NWFZ which has come into force in 22 countries of Latin America.

Non-aligned countries which have renounced nuclear weapons have pressed for assurances from the NWSs that nuclear weapons would not be used against them. At UNSSD I the five NWSs made separate declarations concerning non-use. In 1979 the Committee on Disarmament set up a working group to consider and negotiate on effective international arrangements to assure NWSs against the use or threat of use of nuclear weapons. So far it has not proved possible to reconcile the different assurances into a common undertaking, for example in the form of an international convention or a Security Council resolution.

Other Weapons of Mass Destruction

Apart from nuclear weapons, three other categories of weapons of mass destruction were identified by the UN in 1948: biological, chemical and radiological. For some years the first two were grouped together, since the use of both types in war had been banned by the Geneva Protocol of 1925. Negotiations made little progress until in 1968 the UK proposed that biological weapons should be dealt with separately, on the grounds that they had never been used in war, there were no

The Case for Negotiated Disarmament

large stockpiles requiring elaborate verification procedures, and the effects on humanity of 'germ warfare' would be far more horrible and unpredictable than the use of the nerve agents.

The British initiative led to the abolition of existing stockpiles under the Biological Weapons Convention of 1972. The UK supported Swedish efforts at the Review Conference in 1980 to improve the procedures for ensuring compliance with the convention, but no progress could be made because of Soviet opposition. The Soviet Union has also resisted requests for a convincing explanation of an outbreak of anthrax which took place in the city of Sverdlovsk in 1979.

The present priority is a similar ban on the production and possession of chemical weapons – the nerve agents usually delivered in the form of poison gas. The UK, which has no offensive chemical-warfare capability, tabled a draft convention for total prohibition in the Conference of the Committee on Disarmament in 1976. Bilateral discussions between the US and USSR began immediately afterwards. They have made some important progress on the elements of a treaty, but verification remains the major obstacle. Efforts have therefore been made in the Committee on Disarmament to encourage more openness and to build greater confidence in the effectiveness of a convention. Particularly useful work on verification procedures was done in the chemical warfare working group under Swedish chairmanship during 1981.

Radiological weapons have been defined as those which would be designed to spread contamination, without a nuclear explosion by indiscriminately releasing radioactive materials stored in the weapon into the environment. In the opinion of experts such weapons do not currently exist, but it would be a useful step to block off this area of military development. In 1979 the US and USSR tabled a joint proposal to ban radiological weapons, and a draft treaty is now under negotiation in the Committee on Disarmament.

Conventional Weapons

A preoccupation with nuclear weapons in the international disarmament discussions has tended to divert attention from the more immediate threat of conventional weapons. A British proposal[20] at UNSSD I for a study on ways of limiting the world-wide conventional arms build-up failed because the major developing countries are opposed to any measure which they believe might interfere with their right under the UN Charter to acquire weapons to defend themselves,

134 *The Case for Negotiated Disarmament*

and because the Soviet Union sees political advantage in a world of conflict as part of the 'revolutionary process'.

In 1980 Denmark reintroduced the idea of a study on all aspects of conventional forces and weapons. A resolution[21] endorsing this approach was adopted at the 35th General Assembly, despite opposition from the Communist bloc, India and Brazil. The Soviet Union and India combined to block agreement on the terms of reference of the study at the meeting of the UN Disarmament Commission in 1981, so it will now be delayed for another year.

There is a particular class of conventional weapons which has always been a matter of special concern: the 'inhumane' weapons which cause unnecessary suffering or are indiscriminate in their effects. At the UN Conference on Inhumane Weapons in 1980 new prohibitions were finally agreed. The UN Convention adopted[22] is based on a joint Anglo-Dutch draft of the legal framework, to which three protocols are attached placing restrictions on the use against civilians of mines, booby-traps and incendiaries (including napalm), and an outright ban on weapons producing fragments not detectable by X-ray. The Convention was opened for signature in April 1981. It comes into force when 20 states have ratified it.

In Europe, the control of conventional forces is being discussed at the Vienna talks on Mutual and Balanced Force Reductions (MBFR). There is general agreement that the objective of the negotiations should be to reach a common collective ceiling which establishes parity in the ground force manpower of NATO and the Warsaw Pact in Central Europe. The main obstacle to progress is the disagreement between East and West over the size of the Warsaw Pact forces in Central Europe. Western figures indicate that the East enjoys a substantial superiority. Continued efforts are being made to resolve this problem, so that the participants can move towards the conclusion of a first-phase agreement.

At the CSCE Review Meeting which began in Madrid in November 1980 the Western side continued to support the French proposal[23] for a Conference on Disarmament in Europe to negotiate binding, militarily significant and verifiable confidence-building measures applicable to the whole of Europe, including the European part of the Soviet Union. In February 1981 President Brezhnev conceded the territorial application of confidence-building measures, but did not reveal the Soviet Union's attitude to the other French criteria. Measures which met these conditions would make a valuable contribution towards reducing tension and the danger of armed conflict in Europe.

The Case for Negotiated Disarmament

Advances in Science and Technology

In an age when so much of the world's scientific research effort is being channelled into the military field, there is a constant danger that it will give rise to developments in weapons technology that outstrip progress in arms control. Lord Zuckerman drew attention to this phenomenon in a lecture[24] to the American Philosophical Society in November 1979, and suggested that governments would do well to control the in-built momentum of work in the weapons laboratories on new and more dangerous systems. Proposals have been made from time to time for placing general constraints on military research and development, but nothing has come of them for the simple reason that the secrecy of sensitive security work is jealously guarded and any agreement on this aspect of military effort would be completely unverifiable.

There is one specific example of constraints in new weapons systems in the 1972 ABM Treaty. Under this US/Soviet agreement the deployment (but not the development) of advanced devices such as lasers in an ABM role is banned. The treaty states that before such devices could be deployed for ABM purposes the two parties would be obliged to consult. But the work of Soviet scientists on laser-damage weapons has recently given rise to an increase in research and development funds allocated for this purpose in the US defence budget. The treaty itself comes up for renewal in 1982.

Another approach lies within the 1967 Outer Space Treaty, which contains provisions aimed at stopping testing any types of weapons on celestial bodies. Nevertheless outer space provides a good example of science and technology being perverted for military ends. Recognising that the treaty did not actually forbid military activities in space, other than on the moon and other celestial bodies, the Soviet Union carried out tests on a hunter-killer satellite between 1968 and 1971, again between 1976 and 1978, and in 1980-1. Their objective was to develop a military capability to destroy other countries' satellites used for information gathering, navigation, communications, command and control, and conceivably to destroy satellites used for monitoring arms control agreements — although the last application would violate the provisions of SALT I. The development and potential use of such a capability would deal a crippling blow to an opponent's deterrent system and give the possessor a crucial strategic advantage.

In 1978 the United States decided that, in view of the Soviet advances in anti-satellite warfare (ASAT), they would have little choice but to enter the space-weapons defence field. But their preferred course was to seek negotiations with the Russians on arms control in outer space, and bilateral ASAT talks began in June 1978. Some progress was made, but in June 1979 the Soviet Union rejected a US proposal for a one-year ban on the testing of anti-satellite weapons. The USSR also asked that the US Space Shuttle be included in any ASAT ban, which the US did not accept. It is unlikely that any agreement will be reached in the near future because there are major issues still to be resolved, such as how to verify an agreement and whose satellites to include.

Scientific advances can also be used to assist the disarmament process, for example in satellite observation. At UNSSD I France proposed the establishment of an International Satellite Monitoring Agency (ISMA) under UN auspices to help in the verification of arms control agreements in the same way that the superpowers monitor each other. An experts group was set up to study the technical, legal and financial implications. This group has now reported, and the subject will be pursued at UNSSD II in 1982 under the heading of future disarmament institutions. One of the principal problems will be the cost of establishing and operating an ISMA; others include the choice of satellite tracks, the processing of the data acquired, and its dissemination to UN members.

Obstacles

The nuclear arms race, like competition in any other area of military technology, is fuelled by the fear of falling behind a political opponent in some crucial aspect of military power. It is rooted in the political differences between East and West. There are therefore enormous problems involved in devising and implementing practical measures of arms control and disarmament acceptable to both sides. Some of the difficulties spring from the very nature of open and closed societies. Things which are taken for granted in the West — such as the publication of detailed information on defence budgets and armed forces, and public debate over new weapons systems — are quite unknown in the East.

The Soviet attitude is critical to progress because of its relation to the three criteria adopted by the United Nations for all arms control and disarmament agreements: verifiability, balance and undiminished security. Whereas Western states have offered to accept all kinds of verification measures (including on-site inspection) necessary to ensure

The Case for Negotiated Disarmament

compliance with arms control agreements, the Soviet Union has traditionally referred to international inspection as 'legalised espionage'. As long ago as the World Disarmament Conference in 1932-4 the Russians were rejecting international supervision of disarmament.

Balanced disarmament sometimes requires asymmetrical reductions. But there is little incentive for the Soviet Union to negotiate in areas where it has built up a striking military advantage (for example in long-range TNF, chemical weapons and conventional forces). Consequently there is pressure on the Western side to catch up in order to provide leverage in the negotiations. As for security, the countries of Western Europe are understandably reluctant to accept one-sided proposals which would give the Soviet Union a power of veto over their defensive arrangements. With the examples of Hungary, Czechoslovakia and Poland constantly in mind, they see themselves as becoming increasingly vulnerable to Soviet political pressure and blackmail if they were to allow a further deterioration in their defensive capability in relation to the Warsaw Pact.

It is not surprising that Soviet efforts in the field of 'disarmament diplomacy' are concentrated on the goal of achieving Western disarmament. The latest of these is aimed at stopping the deployment of new US missile systems in Europe and at preserving the existing Soviet advantage in theatre nuclear forces. No effort or expense is spared in directing propaganda at West European countries with this objective. A US Congressional Committee[25] has estimated that $200 million was poured into this and other special campaigns in 1979. Another fruitful area of propaganda is the Soviet claim to be the world leader in putting forward disarmament 'initiatives' — a claim which does not stand up to close examination.[26]

But some Western observers[27] have speculated whether the Soviet Union may be nearing a point at which genuine disarmament becomes an economic necessity. The burden of military expenditure puts an increasing strain on the Soviet economy and distorts the allocation of resources. According to a compilation from an independent source,[28] the Soviet Union's military expenditure in 1977 was 1½ times its spending on health and education combined. The enormity of this burden can be seen by a comparison with the UK, where the defence budget was less than half of health plus education.

Maintaining international peace and security is the primary purpose of the United Nations, as stated in Article I of its Charter. Over the years continuous efforts have been made to encourage the use of UN machinery for peace-keeping and the peaceful settlement of disputes.

138 *The Case for Negotiated Disarmament*

The Memorandum on UN Peace-keeping Operations[29] put forward by the UK and its Western partners in the Special Political Committee in 1978 was accepted by most states, but not by the Soviet Union. As a Permanent Member of the UN Security Council it has the power to block the deployment of peace-keeping forces by its veto. Whereas the British contribution to UN peace-keeping in 1977 was $25 million, the Soviet Union provided only $9 million.

Soviet Motivation

Various reasons have been put forward for the reluctance of the Soviet Union to accept genuinely reciprocal measures which would place constraints on its own military power. Those who start from sympathy for the heavy losses sustained by the Soviet Union in the Second World War regard it as natural for the Soviet leaders to over-insure against any possible repetition, and to fear encirclement by hostile powers, with the possibility of having four adversaries with nuclear weapons in any conflict. Some even defend, with reference to such fears, the Soviet Union's continued concern to dominate neighbouring states in the form of a buffer zone and to keep control over its unreliable allies in Eastern Europe, and the obsession with secrecy in military matters which makes it impossible to publish information on its armed forces or open them to outside inspection.

Others stress the place of military might in maintaining the Soviet Union's status as a superpower, with the result that the military exerts an unusually large influence. Others again point to the totalitarian nature of the Soviet state, in which there is no independent public opinion, still less any political opposition, to put pressure on the leaders to negotiate or reduce military expenditure.

All these factors probably play some part in the decisions of the Soviet leadership; but the most important is the belief in the 'world revolutionary process' of Marxist-Leninist doctrine, which it is the Soviet Union's duty to promote, and which will inevitably lead to peace, because 'socialism and peace are inseparable'.[30] Combined with the tradition of imperialist expansionism going back to Tsarist days, this view of the world represents a formidable barrier to progress, and is the main reason for the Soviet unwillingness to be associated with international control.

In statements for internal consumption the Soviet leaders, though denouncing anything they construe as a Western attempt to negotiate from strength, derive satisfaction from their own military supremacy and their conviction that the 'correlation of forces' is shifting in their favour. This view was recently expressed in an authoritative article[31] in a Moscow journal dealing with the ideological struggle. The writers restated the Soviet position that the spread of communism must be encouraged as a force for world peace and progress. It is a view not shared by the rest of the world. The Soviet military occupation of Afghanistan was condemned by 111 states at the 35th UN General Assembly as a threat to international peace and security.[32]

President Reagan has made a more direct linkage between Soviet aggression and the prospects for arms control agreements. At the same time, Western states recognise that it is not a question of offering arms control concessions to the USSR as a reward for 'good behaviour'. The need is for balanced arms control measures which will enhance all-round security and make both sides feel safer. To achieve this there has to be a high degree of unanimity on the Western side, supported by pressure from the non-aligned countries in areas where the chief obstacle is Soviet intransigence, for example in such matters as openness and transparency in military budgets. Fortunately there are indications that the non-aligned majority at the UN are becoming increasingly aware of these factors and less susceptible to Soviet propaganda. There is a possible alliance of interests here which could be harnessed both in the Committee on Disarmament and at UNSSD II in 1982.

Prospects

Some critics of arms control contend that the negotiations have failed and other solutions must be found to curb the arms race. Others urge a different approach to the problems of international security. But the fact remains that, in a world of over 150 sovereign states, negotiated arms control is the only practical way of enhancing international security; the two are inextricably linked. It is also the only way to fulfil the three criteria of verifiability, balance and undiminished security for all states.

The results achieved in the negotiations may appear pitifully small, but their significance should not be underestimated. Already the nexus of existing bilateral arms control agreements, together with a number of multilateral agreements such as the Partial Test Ban and Non-Proliferation Treaties, circumscribe a variety of activities in the

140 *The Case for Negotiated Disarmament*

military nuclear field. In the East/West context, it is relevant to speculate about the heights of unbridled competition in strategic arms which might have been reached without the constraints of the SALT process.

But progress has been frustratingly slow, because of the vital security interests at stake. Much remains to be done, not only to curb the strategic arms race but to limit theatre nuclear forces on both sides, to ban nuclear weapons tests, to strengthen the non-proliferation regime, to abolish chemical weapons, to reduce the appallingly high level of world military expenditure, to introduce militarily significant confidence-building measures, and much more.

It would be totally wrong to let frustration at the difficulties lead to a headlong rush into profoundly dangerous and destabilising policies, such as unilateral disarmament. The most sensible way to proceed is to assume that East and West can find shared interests — after all, the cost of armaments is a burden to Eastern and Western economies alike. But arms control is not a soft option; it calls for a hard head, realism and patience. It calls for a renewal of the political climate for superpower negotiations which was soured by events in the 1970s. It calls for a willingness on the part of the international community to take action to implement the recommendations of UNSSD I.

The prospects are admittedly not bright; it would be over-optimistic to expect any major treaties to be concluded before UNSSD II in 1982. It would also be unrealistic for UNSSD II to set up a timetable for the conclusion of future agreements in the present atmosphere of tension and mistrust. The comprehensive programme of disarmament now being worked out in the Committee on Disarmament[33] will be an important document in charting the future, as will the results of the major UN disarmament studies due for completion in 1981.

Preparatory meetings for UNSSD II began in New York in May 1981. From the memoranda already sent to the Secretary-General, it is clear that many countries see a major function of UNSSD II as a review of developments in arms control since 1978 and a post-mortem on what has impeded progress. There may also be a tendency to move away from the global approach and to put greater emphasis on regional arms control, based on the recognition that the various continents have differing security problems. Western states have argued for a better balance between nuclear and conventional arms control, on the grounds that the greatest threat to international security may come from the continual conflicts with conventional weapons taking place in areas where nuclear deterrence does not operate.

The Case for Negotiated Disarmament 141

What is badly needed is a general realisation that multilateral arms control measures, endorsed by all member states of the United Nations, can be just as important as the acquisition of military strength in preserving national and international security and bringing to an end the conflicts which could threaten the human race with extinction. Negotiation is the key to survival. But for efforts to be successful, all states must approach the negotiations seriously, dispense with declaratory rhetoric, and get down to the real business of working out the terms of concrete agreements.

Notes

1. *Statement on the Defence Estimates 1981* (HMSO, London, April 1981).
2. Text issued by Lambeth Palace in December 1980.
3. Neville Brown, *A British Approach to Peace* (Fabian Society, London, 1981).
4. E.P. Thompson, *Protest and Survive* (Penguin, London, 1980).
5. Major-General A.S. Milovodov and Dr E.A. Zhdanov, *Voprosy Filosofii*, no. 10/80 (Moscow, 1980).
6. Text published in *Arms Control*, vol. 1, no. 2 (Frank Cass, London, 1980).
7. *New Statesman* (London), 24 April 1981.
8. See *The United Nations and Disarmament 1945-1970* (UN, New York, 1970).
9. *The Military Balance 1980-81* (International Institute for Strategic Studies, London, 1980).
10. *World Military Expenditures and Arms Transfers 1969-1978* (US Arms Control and Disarmament Agency, Washington, 1980).
11. *World Armaments and Disarmament: Stockholm International Peace Research Institute Yearbook 1980* (Taylor and Francis, London, 1980).
12. *World Military Expenditures and Arms Transfers 1969-1978*.
13. *Against Nuclear Arms and War* (World Disarmament Campaign, London, 1980).
14. UN Document A/35/392, 12 September 1980.
15. See Nicholas Sims, *Approaches to Disarmament*, 2nd edn (Quaker Peace and Service, London, 1979).
16. UN Document A/RES/S.10.2, 30 June 1978.
17. UN Document A/AC.187/96, 1 February 1978.
18. UN Document A/34/42, 25 June 1979.
19. *World Armaments and Disarmament: SIPRI Yearbook 1981* (Taylor and Francis, London, 1981).
20. UN Document A/AC.187/96, 1 February 1978.
21. UN Document A/RES/35/156A, 12 December 1980.
22. UN Document A/RES/35/153, 12 December 1980.
23. CSCE Document RM.7, 9 December 1980.
24. Lord Zuckerman, *Science Advisers, Scientific Advisers and Nuclear Weapons* (Menard Press, London, 1980).
25. *Soviet Covert Action* (US Government Printing Office, Washington, 1980).
26. See Roy Dean, 'Disarmament and the Soviet Union', *The World Today* vol. 35, no. 10 (RIIA, London, October 1979).

27. See Peter Hennessy, 'Russian Leaders Face Some Sharp Choices', *The Times*, 14 April 1981.
28. Ruth Leger Sivard, *World Military and Social Expenditures 1980* (British Council of Churches, London, 1980).
29. UN Document A/RES/33/114, 18 December 1978.
30. President Brezhnev at the 25th Congress of the CPSU, Moscow, February 1976.
31. Yuri Zhilin and Andrei Yermonsky, 'Once More on the World Balance of Strength', *New Times*, no. 46 (Moscow, November 1980).
32. UN Document A/RES/35/37, 20 November 1980.
33. The Western working paper CD/205, 31 July 1981, tabled by Australia, Belgium, Germany, Japan and the UK, is an example of a phased and co-ordinated plan of action.

7 INTERNATIONAL HUMANITARIAN LAW: PRINCIPLES AND PRACTICES

Geoffrey Best

This is a chapter of counterpoint. The rest of the book has been about the theory and, so far as it is understood, the practice of nuclear deterrence. The 'practice' side of it evidently poses a peculiar problem. In any other discussion of strategy and weaponry theory and practice, the former can be checked against the latter in terms of results; for example 'Britain imposed total blockade on Germany in 1914-18, with such-and-such consequences . . . ', or 'Germany towards the end of 1939-45 resorted to aerial bombardment by its V1 and V2 weapons, with these apparent results . . . ' The calculation of such results defies precision, not least because they were not achieved in isolation from concurrent belligerent activities; but enough figures of damages, deaths, losses of production, territory and even morale, etc., are normally available for estimates to be more or less well-informed. The theory can be checked against the consequences of the practice, and judged accordingly. With the so-called nuclear deterrent, however, consequential evidence is largely lacking and most of it is speculative. There is disagreement as to how much it has actually 'deterred', and even whether it has deterred at all — some observers doubting whether it really is the main cause of the nuclear-armed powers not having gone directly to war with each other since 1945. Figures and calculations abound in the discussion, but in the nature of the case they relate to little concrete experience. We may be able to imagine — which is to say, riskily, that we tell ourselves we can imagine — what nuclear war must be like, but, the events of 6 and 9 August 1945 and subsequent atmospheric tests apart, our estimate of 'results' is speculative, and, in the strict sense of the word, imaginary. The discussion is of a kind of war which has not actually happened and which the discussants are primarily interested in preventing from happening.

Salutary counterpoint to all that is the recollection that since 1945 wars of more familiar kinds have not ceased to happen, and that about their results nothing at all is speculative or imaginary. They include something between 10 and 20 million deaths. Greater precision is made impossible by the nature of most of these wars, which has been to victimise above all non-combatants, and either to cripple their normal means of subsistence — poor enough for most of them at the best of times — or to turn them into refugees, with all that that implies by way of misery, sickness and extinction. When war works like that, the fall-out area of its effects becomes almost immeasurably enlarged, and exact accounting of its victims proportionately more difficult. Armed forces of whatever sort usually keep a close tally of their own losses, but only the more efficient and responsible of governments can do as much, if they should care to do as much, for civilians. The exact extent of loss and damage due to these so-called 'conventional wars' therefore also defies calculation, but only because there has been so much of it. If, nevertheless, it should still be difficult for contemporary Western Europeans to imagine, as one senses that it is, that cannot be because it has not actually happened to people somewhere else, but more because it is so different from our common understanding of what war is, and what it should be like.

The modern European and North American experience of war has been, by world standards, both peculiar and significant. Peculiar, because so much of it has for so long been confined to the operations of 'conventional', more or less 'regular', and professionally-run armed forces against each other, rather to the exclusion of civilians; significant, because the same process of cultural and political development which thus moulded European states' mutual hostilities produced *pari passu* a code of laws and customs to regulate and — such were the intention and hope — so far as possible to civilise them: the international law of war, peace and neutrality. That code, formed in Europe, has within the past hundred years been stretched to serve the world, and in its contemporary form, the international law of armed conflict, provides the society of states' only common criteria for judging the rights and wrongs of the conduct of wars and (more vaguely) common principles for appraising their justifications. Wars are now declared to be 'just' or 'unjust' as readily in South East Asia or the Muslim-Arab world as in that of the North Atlantic (*more* readily, indeed), but the language derives from modern Europe's Romano-Christian foundations. (For more on this, see Chapter 2 above p. 41.) When atomic and thermo-nuclear bombs are judged 'unlawful' and atrocious, in the same way that other means of waging war have been so judged from time to time, the formulation of the indictment is in terms of the classic 'law of war' raised on those foundations. Prohibitions and restraints in war are to be found in most cultures, past and present, but the European code of them happens to be the one from which the world has drawn its language about them.

Theory is one thing; practice — we come back to it — another. This law of war was evolved by Europeans to make their 'deadly quarrels' less damaging, morally as well as materially. It has always been and it remains a moot question, whether it succeeded in that aim. The question is, strictly speaking, unanswerable, so many variables having to be taken account of and so much of what is relevant being imponderable. Nevertheless many might agree to share the belief of the International Red Cross movement that (until 1939, anyway) Europe's wars, though often cruel and reckless enough with the law, would have been even more so without it. The war of 1939-45, in many aspects unprecedented, apparently marked a turning-point; but in which direction and to what extremes the practice of war in Europe could then have gone must remain unknown, hardly any wars having happened in Europe since then. Theory's relation to practice cannot be answered out of that laboratory. Other parts of the world, however, more than make up for Europe's passivity. The international law of war, its pre-1914 formulations somewhat modernised between 1945 and 1949 (especially in the four Geneva Conventions of that last year) to meet the needs of a new era, was available to work its softening effects upon all the wars which soon began to happen in Asia and Africa (and to a much lesser extent Latin America), and was indeed all that was available for that purpose. The question therefore becomes as difficult and delicate as it can be; not whether a European-based body of international law can work in Europe, but whether it can work elsewhere.

The theory of international law is rooted in belief in the existence of certain universal utilities and affections. Any international law purporting to be also universal must rest on plausible assumptions that there is nothing in it to which any party cannot, and much to which all parties can, subscribe. Europe's international law was not born universal. Universality was thrust upon it by the European and North American commercial empires which took it round the world with them, relied upon it for the conduct of their own interrelationships, and by natural processes imparted it to their political protégés and trading partners, from China to Peru. European international law became the world's, but it always was and it has remained an open question, whether this was simply 'accidental', through force of circumstances merely, or because it contained objectively universal principles which fitted it ideally for the purpose.

The question could not be much mooted so long as Europe and the United States evidently ruled the roost, as they did until 1945. German ultra-nationalist graspings towards a major revision foundered in her national catastrophes; the USSR registered many objections but nevertheless pragmatically accepted what parts of the law it needed for practical purposes; and in any case the USSR, for the twenty years or so after the October Revolution, remained something of an outsider and an anomaly in world affairs. The Second World War and its aftermath changed all that, to an extent and in a manner which it would be otiose to sketch here. The countries which gave the world its international law no longer have the preponderant power in it they used to have, and every branch of that law is open to scrutiny and questioning in countries which may find cause to feel that its original cultural character biases it against them.

The international law of war has been thus questioned, for example, by two largely coinciding clusters of opinion: 'national liberation' fighters against (to use the language which had become standard in socialist and 'Third World' circles through the 1960s) 'colonial domination and alien occupation and racist regimes'; and their most aggressive established-state backers and spokesmen at the Geneva conferences of the 1970s to update the Geneva Conventions, North Vietnam and the People's Republic of China. The former, broadly speaking, alleged that the cards, having been dealt by states (and empires) with conventional armed forces designed to keep them in being and with political philosophies tending to justify the *status quo*, were stacked against 'irregular' and guerrilla combatants with equal justification for overturning it. China and North Vietnam on occasion let ideology carry them to the point of maintaining that every military enterprise of states other than (self-defined) socialist people's republics was morally and legally damned from the outset, while anything done against them by (self-styled) national liberators must by definition be all right. This was so extreme and wholesale a rejection of the inherited system that it found no backers among the national liberators' other usual patron-states or, indeed, among most of the national liberation movements themselves, so far as their representatives at and around Geneva then let it be understood. Far from wishing to challenge it in its essentials and totally to reconstruct this part of international law which affected them so directly, they sought only to repair the parts of it which were working badly. European though its historical origin and moral essentials might be, they were not, apparently, felt to have become generally unacceptable.

This almost unqualified endorsement of the universality of the international law of war was welcomed by the International Committee of the Red Cross, whose articles of long-held faith it cheeringly confirmed. The ICRC, the entirely Swiss-manned curator and executive of international humanitarian law and its particular keystones, the Geneva Conventions, is of peculiar importance to our theme. Quite soon after its foundation in the 1860s it acquired the unique position it has retained and expanded ever since, of representing mankind's conscience concerning war: not necessarily condemning it but recognising its waste, tragedy, pain and shame, and devoting itself to mitigating them. The Red Cross movement at large does something of this sort through the national societies which compose it, but their usefulness in the larger realm is limited by their necessary patriotism. Only the entirely neutral and impartial ICRC can represent the supranational interest of mankind in the limitation and restraint of armed violence between states and within them. Far more than any other organisation or body it, so to speak, 'stands for' the law of war, representing it to and in the world through its general concern about and special interest in the area of law comprising war and peace on the one hand, human rights on the other. It has not sought this remarkable role and position of authority. Its ways have always been quiet and modest; it has to even a self-damaging extent shunned limelight and headlines, and what influence and power it possesses come certainly not from strength, conventionally measured, but from its weakness. How many divisions, Stalin might equally well have asked, has the ICRC? Yet its influence and authority seem to have done nothing but grow in the world since 1939-45, not through ambition but (as indeed was the case with all its earlier extensions of activity) through responsibility. 'Geneva law', that part of the classic law of war covering victims of armed conflicts (prisoners of war, sick and wounded combatants, and non-combatants in general), became increasingly difficult to view or to handle separately from 'Hague law', the part of it codified at the Hague Peace Conferences of 1899 and 1907, the part more to do with means of conducting conflicts — what means may be used to injure the enemy and what respect must be paid to his power by those subject to it, etc. The ICRC therefore grasped the nettle of combining them. Following the reinvigoration of the concept of 'human rights' signalised in the 1948 Universal Declaration and its progeny, and recognising the extent of the affinity between them and its own original proper concerns, the ICRC followed the logic of the situation by admitting an interest in the law relating to 'human rights' and clarifying the nature of its relation to the law of war within the umbrella concept of the international humanitarian law it now regards as its special province.

In this area of law no writer is more distinguished or authoritative than Jean Pictet. Internationally respected for his standard Commentary on the 1949 Geneva Conventions and the doyen of the ICRC's strong team of juridical experts until his formal retirement (after a lifetime's service) in June 1979, he has written with equal ease about *The Principles of International Humanitarian Law* and *The Fundamental Principles of the Red Cross*. These two booklets have almost everything in common. It is clear that, to M. Pictet's mind, these principles constitute not so much two sets as the same set looked at from slightly — only slightly — different angles. A man of philosophical mind as well as religious conviction, he begins with definitions. At the back of the key words 'humanitarian' and 'humanitarianism' is the term 'human'.

> In the meaning which is of interest to us here, 'human' denotes a man who is good to his fellow beings . . . varying degrees [of] . . . kindness, generosity, devotion, faithfulness, tenderness, pity, compassion, a spirit of mercy, gentleness, patience, clemency, toleration, constancy, forbearance, commiseration and others . . . If one wishes to summarize all that and interpret it from the practical point of view . . . one would say the following. Animated by favourable intentions, the good man is touched by the sufferings of others and he tries to alleviate them; showing them respect and affection, he protects and helps them. In other words, he devotes himself to them. With complete equality of mind he suffers misfortune, is not carried away by anger against anyone, and forgives joyfully.[1]

The man whose mind is set like that, says Pictet, may properly be characterised as being guided by a sense of 'humanity'; that is, following Littré, 'a sentiment of active goodwill towards mankind'; in company with which, he affirms, 'pity' and 'charity' will usually be found. 'Humanitarian' action is 'action beneficent to men'; 'humanitarianism', 'the universal social doctrine which aims at the good of all mankind'.[2]

The Christian sources of Pictet's own presentation of these principles are nowhere more to be apprehended, perhaps, than in his frequent reliance on his great ICRC predecessor, the more overtly Christian Max Huber — his glad acceptance of Huber's analogy of 'the Good Samaritan' — and the sensitive exactness with which he distinguishes 'charity', the all-important primary part of goodness, from 'humanitarianism' on the one side, 'justice' on the other.[3] He takes care, however (as a spokesman for such a universal body of principles understandably must), to avoid imputing to 'humanitarianism' anything like a primarily Christian character. A reader ignorant of charity's special meaning for the Christian would not be informed of it by anything Pictet specifically says. There are no acknowledged quotations from the Christian Scriptures. The only obvious and direct quotation from them, 'Whatsoever ye would that men should do to you, do ye even so to them', is not presented as such but as a

> fundamental precept [which] can be found, in almost identical form, in all the great religions ... It is also the golden rule of the positivists ... It is indeed not at all necessary to resort to affective or transcendental concepts to recognize the advantage for men to work together to improve their lot [and to recognize also] the concept of solidarity as an ideal for the organization of the community. The maxim, do unto others as you would have them do unto you, another version of the 'golden rule' cited above, therefore represents a universal truth, for it is in full conformity with human nature and the needs of society.[4]

Humanitarianism, he is at pains to explain, is not a religion (though clearly it has much religious impulse in it). It is something at once larger and smaller than a religion; smaller, because avowedly this-worldly in its vision and purpose; larger, because (following the model outlined by Gordon Dunstan at the start of Chapter 2) it spreads beyond the normal provinces of the self-styled religions and gives some common elements in them the means to become bigger than their single selves.

> Humanitarianism is not a religion opposing itself to other religions, a moral philosophy set up in opposition to other moral codes. It does however coincide with the precepts of many religious and moral codes ...
>
> No longer speaking of malediction striking humanity, of guilt or a fatal destiny, humanitarianism has really overcome one of the chief collective inhibitions from which the world used to suffer. It has preserved from its moral and religious sources only what was rational and universal. Having done so it has in no way tried to deify man. It has merely taken him as the object of its interest. The humanitarian doctrine is therefore one of the rare fields where people of all beliefs can meet and stretch out the hand of friendship to each other, without betraying matters which are closest to their hearts.[5]

It is not coextensive with justice or charity, justice (at any rate in its more familiar forms) being morally less sensitive and charity less calculating; but it derives much of its zeal and drive from them, and wholeheartedly embraces justice in its higher, ideal forms.

> To conclude, the Red Cross movement gathers under its flag all those who wish to serve, even though the deeper reasons for their commitment may differ greatly. As Max Huber wrote, 'The most varied points of view in philosophy, religion and human experience enable man to understand the idea of the Red Cross, the moral principle it embodies and the action it demands.'[6]

The question to be considered in the context of this book is how far this scheme of the universal ethical foundation for international humanitarian law corresponds to reality. Its appraisal must begin with marking the limits within which its ICRC proponents write. Jean Pictet, since Max Huber's death the chief proponent, writes from within the 'Genevan' juridical tradition and is careful not to go outside it. That tradition, until within the last fifteen years, had almost exclusively to do with the protection and salvaging of people from the ravages of war. It had nothing explicitly to do with strategic planning and the conduct of combat. It went off the road of combat with the Good Samaritan, not onward along it with 'the Good Centurion'. There is nothing in it to which the Christian soldier — we may assume that he is fighting with a good conscience for a cause objectively seen to be just — can possibly object; he is, however, likely to observe that there is little in it to help him while he is actually doing the fighting. Not Geneva law but (until the 1970s) Hague law has to be his guide in that respect.

Hague law represents in a more painful form the awkward compromise upon which the law of war has to rest: the compromise between humanitarian principle and military reality. 'Geneva man', if we may so call him, may (by accident, misunderstanding or malice) lose his own life in war but, so far as other people's lives are concerned, he is only there to save them. 'Hague man' carries the peculiar moral burden of being there to take them. He has a right to kill. The Good Samaritan has nothing useful to say directly to him; neither have the Decalogue or the Beatitudes. He dare recall them only through a venerable filter of casuistical or dialectical interpretation. Christians (to speak of the only faith the author knows much about) have over many centuries become expert at this sort of thing, but for the sensitive and conscientious it must always remain embarrassing and worrying, because ideally it demands walking along the knife-edge of a paradox. M. Pictet's principles cannot come into the mind of the good man at war on better terms than coexistence with his perception of the demands of military necessity, not, therefore, necessarily neglected, but at best bracketed with the latter; incapable of assimilation to military necessity, but able, with luck, to survive in tension with it. His old elitist sense of honour was the soldier's principal aid to sustaining that paradox and no doubt still can be, wherever its classic calls to sacrificial altruism and respect for the helpless have not been too much eroded, vulgarised or perverted. Hague law largely relied on the operative power of that sense in its expectation that fighting men would accept this undeniable risk, forgo that apparent advantage and endure the other irksome inconvenience (for example, don't shoot if it might be a civilian, treat enemy wounded as you would your own, leave to civilians some food though hungry yourself) rather than break the rules or lower the tone. There is little religious or moral principle on the peremptory surface of Hague law, however clearly it may be perceived just under the surface of, for example, Article 22 of the Hague Regulation: 'The right of belligerents to adopt means of injuring the enemy is not unlimited'; Article 23e, forbidding the use of weapons, etc. 'calculated to cause unnecessary suffering'; and Article 25, 'The attack or bombardment by whatever means, of towns, villages, dwellings or buildings which are undefended is prohibited'. Their only explicit reference — admittedly, a momentous one — to the ethical sources of such prohibitions and restraints comes where the preamble to the Convention containing them locates 'the principles of international law' in 'the usages established between civilised nations ... the laws of humanity, and the dictates of the public conscience'. Ever since the Russian jurist F.F. Martens persuaded the 1899 Conference, by inserting those words, to make at least to that extent explicit the law of war's humanitarian commitment, the convergence of its Hague and Geneva streams has seemed logically inescapable.

The Christian and other religious or moral warrior need not therefore be thought to inhabit a conceptual realm so far removed from M. Pictet's as his formal preoccupation with Geneva law and particular respect for the Good Samaritan at first suggested. One aspect of the reality to which international humanitarian law has to relate, the reality of war, is after all within its scope. But what of the other aspect, that to which the Genevan jurists were actually referring in their affirmations of its genuine universality? What of the realities of those other religions, cultures and ideologies of the world, and of the perceptions of war and the world normal within them?

To ask these questions in the context of this particular book is to wonder how much trust – that mutual trust between antagonists (actual or prospective), some degree of which is indispensable to any scheme of prohibition and restraint – can survive in cases of conflict between religions, cultures and ideologies as far removed from one another as some in our world now seem to be. The field thus opened up to inquiry being obviously so vast, the limits must at once be stated of what small entries into it will be attempted here. They can be no more than a reminder, first, of the sources of the mutual trust upon which the law of war has traditionally relied: second, a note of certain disparities apparent on the surface of the international Red Cross movement and some of the difficulties inherent in the notion of international humanitarian law.

A soldier in one of Western or Central Europe's armies, during the heyday of our classic law of war (roughly speaking, the eighteenth and nineteenth centuries), if asked what was the ground of his respect for it and his trust in his enemy's respect for it too, would probably have put it down to Honour; the honour of the regiment, corps, profession, race, even simply his honour as a man. His sense of honour, acquired by learning what his role required of him and by adapting to his surroundings, would, once activated, be so strong as to oblige him to do this or that without reflection on rationale or risks. It could, however, be explained to him that the code of honour, though perfectly capable of appearing to be its own justification, was usually revealed by closer inspection to be the operative device for pursuing purposes capable of more mundane explanations: sociological, political, utilitarian and simply sentimental, with reciprocity as the prime mover of them all. (Trust once disappointed is difficult to reconstruct.) Sentiment is in the mixture because popular military culture, in Europe anyway, has usually contained selective but persistent strands of kindliness, compassion and generosity. Professional self-interest can often be seen at work. Dedication to expertise at killing and avoiding being killed can very well coexist with desire that the enemy, if he takes you prisoner or comes across you wounded on the battlefield, shall treat you decently; he, the fellow-professional enemy soldier, can even become someone whom, albeit normally at a distance, you particularly respect and esteem. The political self-interest of the governments under whom ultimately the soldiers fight can also be discerned. Professional armies and navies were very expensive, to begin with; rulers actually preferred not to have them too severely damaged. But beyond that, their general theory of war was economical and inhibiting. Wars figured

International Humanitarian Law 153

in their codes of law and honour as exceptional, occasional and (in the eyes of some) even unnatural episodes in the general experience of mankind which was felt properly to be peaceful. War, by that philosophy, was to be undertaken only when absolutely necessary, in the last resort and for specific good causes. There was no point in going on with war once its aims were achieved, and there was nothing but harm in letting war go on so long or get so far out of control as to roughen the road to its termination.

The spate of popular national wars introduced by the French Revolution tended not to be of this limited kind but most eighteenth- and nineteenth-century inter-state wars were so, and perhaps their cool, close bonding of theory with practice did not produce worse results for mankind at large than the less limited other sort. Dr A.J.P. Taylor, therefore, had a good deal of history on his side when he remarked, 'Bismarck fought "necessary" wars and killed thousands; the idealists of the twentieth century fight "just" wars and kill millions.'[7] Bismarck's 'limited' wars, fought by highly professionalised and disciplined armies, are worth lingering on. Against the Danes and the Austro-Hungarians, who far from being hereditary enemies were more like brothers, hostilities were kept quite clean. Against Napoleon III's professionalised army in 1870 likewise, honour and legality stayed in the saddle; it served the interests of both parties that the war should appear as a contest between monarchs and states rather than between peoples and nations. A war between peoples and nations is what it nevertheless became after the French emperor's abdication and Gambetta's call to his nation to rise in arms against the odious invader; and very unpleasant it was becoming by the time it ended — lawless, hateful and rather savage. And why did it become so? Mainly because 'the people' had become more involved in it: first, by the direct involvement in the fighting of partisan resistance groups and individuals unlikely to know about the law of war and even more unlikely to care about it; second, politically, by the excitement (not least within Germany) of nationalist passions among the mass public, whose home-bred notion of war more readily included hatred, vengeance and punitiveness than respect, limitation and restraint.

This second phase of the Franco-Prussian War did not last for more than a few months, and the barks just referred to were after all worse than the bites inflicted, so strong were the habitual restraints (for example Christianity, chivalrous military traditions, cultural self-respect) on each side's actual behaviour. The episode nevertheless serves well enough to illustrate the probable fragility of the basis of mutual trust between combatants. Even between the French and the Germans, who had so much in common, it began to go in the winter of 1870/1. On many other occasions before and since then, it has hardly existed at all. The professionally-minded and law-conscious armies which we have so far been considering were European or by natural extension North American; their influence on military organisation and training in other parts of the world has been great, not least because they were the original masters of the modern weaponry technologies which all would-be rivals had to acquire, but indigenous cultural influences in Africa and Asia may be thought to have clearly reasserted themselves within the past two generations. Not much of the values of Potsdam, St Cyr and Sandhurst has appeared in the performance of, for example, the Japanese army 1931-45, General Giap's North Vietnamese army 1950-71, or most of the armies of post-colonial Africa. Professional armies of another kind have been bred in Central and South America, with an eye at least as much on internal security as on foreign warfare, and a quasi-fascist indoctrination in *total* war against the *internal* foe. Here again,though in a very different form, is a military style far removed from the philosophies of Geneva and The Hague. Finally, one may wonder how real can be the understanding of those philosophies held in the armed forces of the Soviet Union and other Marxist-Leninist states. The expression of such curiosity does not, may it be clearly understood, imply that those armed forces are prima facie less law-abiding than others. The day is long past when self-respecting students of war could dare to claim that non-socialist states' armed forces were characteristically innocent of the use of terror against civilians and of the commission of atrocities. So far as discipline comes into the question, there is plenty of evidence to suggest that socialist armed forces observe it better than some famous 'bourgeois' ones. But still the questions may reasonably be put, first, whether the Marxist-Leninist view of the world and of social relations within it must not mean a specifically total and determined sort of war, once so regrettable a thing as war has occurred; and second, whether the Red Cross's view of 'man' as in himself non-ideological — 'just a human being' — can be wholly shared by subscribers to an ideology which denies that 'man' can be seen in (it is argued) so naive and unsociological a way.

Mention of the Red Cross brings us now to the second sector of this review of the principles of international humanitarian law: a glance at some evidence from the national Red Cross societies. Although these societies are the channels by which the international Red Cross idea becomes known to most people around the world, it is essential to remember, first, that the pure milk of that idea is to be found neither in them, nor in the League which provides their ordinary (peacetime) framework, but in the International Committee; and second, that over them the ICRC has little influence and less authority. The doctrine expounded by M. Pictet is the doctrine perceived in the moral and intellectual centre of the movement, the doctrine which *ideally* all its other parts should embrace and to which the ICRC does what it can to make them adhere. Its task therein is at least very difficult, as M. Pictet recognises in the second of his eight chapters on 'the fundamental principles of the Red Cross', the chapter on 'Impartiality'. The realistic Swiss guardians of the Red Cross conscience, by definition neutral with regard to each and every armed conflict they seek to mitigate, know all too well how it gets blunted and stunted in coarser climates and how difficult ordinary belligerent nationals find it to think and to feel *'internationally'* about enemies. The best — indeed the only serious — historian of the ICRC since 1945 remarks in his chapter on its relations with the Red Cross movement generally that 'it is a universal problem to get members of national societies to treat "enemies" humanely . . . It is even more difficult to get intranational "enemies" to treat each other humanely.'[8] During wars, the most the ICRC can do in favourable circumstances (that is when governments are really willing to discharge their treaty obligations) is to persuade the national societies of belligerent countries to help it monitor observance of the Geneva Conventions and to facilitate such good works as they might wish to perform in accordance therewith. Its power of influence in peacetime may not be much greater. The integrity and humanitarian homogeneity of the League is supposed to be assured and protected by the requirements that national societies admitted to the League must also be 'recognised' (that is approved) by the ICRC, and by the theoretical possibility that recognition may be withdrawn from the hopelessly unworthy. The threat of such withdrawal seems to have worked once,[9] but the ICRC's policy has always been to bring and to keep as many national societies within the network as possible, on the very understandable and morally respectable ground that even only a little bit of Genevan influence is preferable to none at all. But that little may never become larger than a government likes to permit, or than the moral imagination of its people can comprehend. Within the Red Cross movement the extent of these national differences has been frankly recognised; they figure quite openly in the extensive operation of self-scrutiny and self-criticism which issued in some of its six 'Background Papers' published in 1975, commonly known after that inquiry's director as 'the Tansley Report'.[10] Better universality than purity, is one way M. Pictet puts it when he touches on this difficult matter: *'fortiter in re, suaviter in modo'* is another.

> The National Societies are the auxiliaries of the public authorities, whose full support they need and with whom they must have relations of full confidence. These Societies cannot exist as foreign bodies within their nations, as Max Huber once remarked . . .
> On the other hand, what we do expect of the Society is that it will remain vigilant and on every occasion will seek to obtain a better understanding of the profound significance of the Red Cross . . .
> The important thing is to remain dedicated, come what may, to the ideal and spirit of the Red Cross. In this domain, we may very well display our intransigence. This ideal and this spirit have been expressed in the substantive principles which . . . rank higher than the others. These the Red Cross cannot surrender at any cost. It will remain faithful to them or it will not survive.[11]

Governments' attitudes are one thing, popular attitudes are — at least in their origin — something else. Another obstacle to the observance of international humanitarian law may be its non-correspondence with regional or 'tribal' styles and standards of behaviour. Most conspicuous but least common examples of this are when a government, without being unrepresentative of the people it speaks for, as an unpopular dictatorship may be, adopts a frankly neglectful attitude to the norms respected by all the others; for example, again, the militarised Japanese government of the Second World War, wilfully dismissive of so much of the ordinary law of war, or more recently the religious-fanaticised governments of Libya and Iran, jettisoning so many of the norms of diplomacy and polite international behaviour. More common by far are occasions when governments, wishing to adhere to their treaty obligations and attempting to enforce respect of them upon their own people, have run into popular misunderstanding and rejection. This may exist equally among civilians or soldiers. Neither, for instance, may be able to understand the law regarding prisoners of war. Incidents when civilians have maltreated wounded troops left behind by retreating armies and have lynched survivors from crashed aeroplanes are as commonplace in the history of war as when soldiers in particularly lawless or tough-talking regiments have practically refused to take prisoners at all.

The question, however, is more delicate in respect of soldiers, they ostensibly being better disciplined than civilians and closer to the sources of official instruction. Civilians' disregard of the Geneva Conventions, etc. must be less surprising than soldiers', of whom so prominent a part of their professional code of conduct is obedience to orders. Soldiers, however, may disregard them for all sorts of reasons. Sometimes officers may wish to control their men but be unable to do so. It was almost always thus, for example, in earlier times when long-defended fortress towns yielded at last to formal assault; officers knew that the troops which survived such risks and terrors would be in such a state of excitement, fury and lustfulness that it was simply impractical to try to prevent, for the first 24 hours or so, the mayhem and looting which had become those troops' traditional reward. Something very similar happened in the last months of the Second World War when the Free French General de Lattre de Tassigny ordered the inhabitants of the south German town which his north African troops had just taken to clear out, leaving their doors and cupboards unlocked, and to stay so many kilometres distant until the evening; his men expected to be able to loot and it was better for all parties to get it over without the addition of rape and murder.[12] When the Red Army's excesses were brought tactfully to Stalin's attention about the same time, he excused them with the classic war lord's argument that brave men who had endured so much deserved some recompense.[13]

Such licence, rooted in military tradition and to some extent explicable in terms of battle experience, is not difficult to understand. The legalist and humanitarian may find it lamentable but they need not despair: there is perhaps little about it which better education and tighter discipline might not cure. More intractable by far is the case of cultural incomprehension. The English master-essayist 'Saki' (H.H. Munro) put his finger on this when he caused one of his characters about 1910 to remark: 'whenever a massacre of Armenians is reported from Asia Minor, everyone assumes that it has been carried out "under orders" from somewhere or other; no one seems to think that there are people who might *like* to kill their neighbours now and then'.[14] How universally appealing after all *is* the Good Samaritan? How many other religions besides Christianity enjoin love of enemies and regard it as evidence of moral strength, not weakness? To the intensely familial and tribal ethics of Asia and Africa (to look no further) may not generalised love of mankind seem a tepid escape from direct responsibility? From some cultural points of view, cannot the infliction of pain on the wicked appear as a duty? And so on . . . It would take a different writer from the present one — a writer well grounded in social anthropology and comparative religion, for a start — to handle this delicate theme with the care and attention it deserves, and the difficulty of handling it properly is perhaps indicated by the fact that hardly ever is notice taken of it in books about the law of war. The striking exception is *The Law of War*, edited by Richard I. Miller, a jurist whose large experience includes work wirh the US Army's legal branch. Part III of this most useful book is on the 'Application of the Law: Russia, China, and Other Powers'. The 'other powers' are a selection from every continent, written about by 'persons with deep knowledge and understanding of the traditions and values of the nations in question'.[15] The acid test they apply is the treatment likely to be given to prisoners of war; would the Geneva Conventions be observed, and if not, what would probably happen instead? A few quotations will convey the gist of this remarkable exercise. Of 'Eastern Europe', for example:

> The cultural differences between populations within such states are apt to be greater than the political differences between states and may strongly color the treatment of prisoners in armed conflicts . . . In those parts of countries where an aristocratic heritage remains, the wounded and sick are likely to be given the same medical care as that which is given to the troops of the detaining power. However, in many communities industrialisation has broken down the previous standards of a rural morality and no humanitarian ethic has taken its place . . . Above all, ethnic enmities are paramount . . . In East Germany the one Western European country in the Communist bloc, local custom will not be important. The camps will be run in accordance with official policy whatever it might be. In the other nations the attitude of the local population toward the country on which the prisoner depends is likely to determine his treatment in the internment camp.[16]

Of 'the Indian Subcontinent':

> India is one of the few nations of the world that might welcome a protecting power to fulfill its convention obligations in international conflict. A principal reason for this is that the role of the impartial intermediary, or arbitrator, is central in the adjudication of domestic legal disputes . . . The Pakistanis are somewhat less enthusiastic about the role of the disinterested neutral, but no less so than, say, the United States.[17]

Of 'Indonesia and Malaysia':

> Literal application of the provisions of [the Geneva Conventions] would be at variance with the Indonesian heritage. Their reaction to the disabled is influenced by Javanese culture, in which the greatest courtesy that one can extend to a fallen enemy is not to inflict further pain and suffering . . . The concept of collective guilt is firmly implanted in the culture. Moreover, the humanitarian prohibitions against torture, humiliation and degradation of prisoners are completely at odds with the cultural tradition of Malaysia and, particularly, Indonesia. Southeast Asians have developed humiliation and degradation of the person to a fine art.[18]

Of 'Zaire, Nigeria [sic] and Central Africa':

> A certain degree of torture, coercion, and degradation of prisoners of war is likely in Central African conflicts . . . Simply stated, an enemy is an evil person . . . Torture would never be arbitrarily imposed, but to the Western mind, the African reasons for torture or degradation are likely to appear arbitrary or even capricious . . .
>
> In practice, there is a total disregard of the Geneva Conventions throughout Central Africa. In large part this is based upon the lack of knowledge of the conventions; but even where knowledge exists, the underlying legal presumptions of the conventions make no particular sense in African terms of reference. The central theory of the conventions is based upon territory and nationality. The African thinks in terms of tribe and religion. Personal humanity and compassion are unrelated to disinterested Western legal philosophy.[19]

And lastly, of 'Latin America':

> The social status of a rebel is critical in determining his treatment as a prisoner . . . Common people should be expected to be interned, tortured, or killed without reference to personal behavior or political position.[20]

The extent and nature of the problem thus boldly outlined, and without pausing to evaluate those particular presentations of it, we may leave it with the reflection that the practitioner of international humanitarian law (the law of war being part thereof) perforce knows how to take such problems in his stride. For one thing he will not abandon hope. On the ICRC's own admission, in Africa, Asia and Latin America it has 'the utmost difficulty in ensuring the application of humanitarian rules, despite their universal acceptance at the conference table'.[21] But people with the Red Cross idea strong inside them will not give up because the difficulties and the dangers mount; they will merely reflect that the risk and sacrifice of their lives may be asked of them more often.[22] Beyond that, they may reflect also that this problem is only one more added to those already inherent in the law of war itself — the paradoxes of reconciling humanity with warfare and of restraining violence in a situation which by definition legitimises it. The first step towards overcoming obstacles must be exact measurement and analysis of them. Law never expects to move anything but modestly *inter arma*, and the global problems sketched by Mr Miller's collaborators may not be greater than they would have seemed on a merely European plane, if sketched by a sympathetic colleague of Grotius three-and-a-half centuries ago. In course of time, our continent learnt how to handle them. So also in time — and change, we know, now happens much faster than it used to, for the better as well as for the worse — may our world.

Such culturally — and ideologically — rooted obstacles to international humanitarian law are one thing; the formal shape and content of the law itself is another, and with a glance at the specifically 'war' part of that we will conclude. Consonance between law and society depends as much on the law's adjustment to social realities as on society's attitudes towards it. The history of the law of war is a cyclical one of repeated disjunctures from social reality and subsequent readjustments to it. Economic and technological change has within the past century or so been the main source of those disjunctures ('bigger and better' weapons, industrialisation, demographic change, the effects of inventions from gunpowder to the laser, etc.), but social and political change have made their mark too: total popular mobilisation, one-party political systems, monocular world views, world revolutionary and national resistance movements and so on have to be brought into the balance if the whole story is to be told.[23]

After the Second World War's tragic demonstration of the many respects in which the law of war was defective, a number of repairs were made; the most striking ones being 'the Nuremberg principles' and the four Geneva Conventions of 1949, extended as they were to cover civilians in occupied territories (a hitherto unprotected category of victims of war), a fair spread of resistance fighters (hitherto an unprotected category of combatant), and to a modest extent *non*-international armed conflicts as well as international ones. Change unceasing, however, soon made the post-1945 repairs seem inadequate. The development of nuclear, radiological, biological and chemical weapons on the one hand, of everything that is supersonic and electronic on the other; the loud grievances and demands of guerrilla fighters in national liberation and other prima facie justifiable conflicts, the dreadful sufferings inflicted upon so many non-combatants during or after them; and the steady growth in mankind's consciousness since 1948 of human rights and every peaceful and progressive thing supposed to attach to them — these were perhaps the main causes of the law of war's coming, by the 1960s, to seem to need another burst of (to use the ICRC's words for it) reaffirmation and development. In the 1970s it got them; another re-working of the Geneva Conventions, this time with the unprecedented novelty of welding with them some major elements of the law of The Hague. These recent achievements are to be found in the two Additional Protocols (additional, that is, to the 1949 Conventions) at last drawn up, at the end of a diplomatic conference spun out over four years, and signed by the 102 states which finished the course in the summer of 1977; and in the United Nations 'Convention on Prohibitions or Restrictions on the Use of Certain Conventional Weapons which may be deemed to be excessively injurious or to have indiscriminate effects', opened for signature on 10 April 1981. Between signature and ratification, however — between the conclusion of negotiations about the most generally acceptable form of words, and the acceptance of the obligation to regulate conduct in accordance therewith — there is a gap which the common reader can hardly be aware of. To sign is one thing; to ratify, another. The UK, for example, did not ratify the 1949 Conventions until 1957. It begins to look as if it may take as long to make up its mind about the Additional Protocols.[24] Already we are (at the time of writing) in the summer of 1981, and still they have not been brought before Parliament and public. Perhaps the UK government wishes to avoid bringing them forward at all. That would be a shame and a tragedy. Indeed they may be open to the charge of 'imperfection' in the very limited sense that no single state's self-interests can be wholly served by a multilateral treaty aiming to protect the victims of war and to restrain its monstrous violence whether that war is between 'undeveloped' and 'developed' or between the 'developed' themselves. It can, however, be argued that the British people ought to be given the opportunity to make up their own minds whether they would wish to become involved in an armed conflict under the 1907-49 rules, or under the 1949 ones as proposed to be amended in 1977.

The 1980 Convention, for its smaller part, is a very modest new instalment of the age-old series of attempts to outlaw weapons which by their very nature are indiscriminate or cause superfluous damage and suffering. (For the UK's admirable share in making it, see Chapter 6, p. 134.) The ICRC has since 1956 sought to bring up to date this branch of the law of war, whose modern origins are to be found concisely in the St Petersburg Declaration of 1868. Chief among the difficulties in its path — given that nuclear weapons are exactly of this most atrocious description — has been the refusal of the nuclear powers to allow those particular weapons to be included in any general discussions. The 1980 Convention has therefore had to be limited strictly to so-called 'conventional weapons'. These, however, can be just as nasty, within their more restricted range and to the immediate victims, as can the nuclear. It may therefore be considered a minor triumph for international humanitarianism that, of the ominously lengthy list of weapons and tactics submitted to the several conferences leading up to the Convention, agreement was reached to prohibit these three: bombs, etc. with non-detectable fragments which, once embedded in a person's body, cannot be traced by X-ray; attacks on military targets from the air with incendiary weapons (for example napalm), if those targets are within a civilian area; and the indiscriminate and unrecorded placing of mines and booby-traps, unless they have self-destructive mechanisms to render them harmless after the passing of the combat period for which they were intended.

It may be remarked that these are rather small beer, these three survivors from the long shopping lists which many participants brought to those conferences. So they are. But if one lesson can be learnt from the history of the law of war, it is that laws and rules placing too great a strain on the self-control, bravery, honour, self-sacrifice or goodwill of belligerents will simply not be observed when the going gets rough. The warrior justly complains when overmuch of those qualities is expected of him, especially when it is expected by people far from the fighting. Having been commissioned to fight, he insists he cannot do it with his hands tied behind his back. We must remark once again that the law of war can never be more than a compromise between what men's humane desires make them wish to do, and what the release of their combative proclivities in war drives them to do. 'Military necessity' has often been pleaded beyond reason (for example and most often, when what was really in question was mere military convenience), but real inescapable military necessity remains an irreducible fact of war none the less, setting limits to what humanitarianism can reasonly demand. A few prohibitions which warring states and soldiers can with goodwill observe may therefore make more sense than a large number which even with goodwill they cannot, leaving the last state of observance worse than the first.

The 1949 Geneva Conventions seeming to some sympathetic observers to go in some respects beyond the reasonable, one of them, the distinguished Dutch jurist B.V.A. Röling, with grim wit remarked that the road to hell might be 'paved with good Conventions'.[25] He was referring particuarly to Articles 33 and 34 of the fourth Civilians Convention, which made bold absolutely to ban the soldier's age-old means of protecting his interests in enemy territory, the taking of hostages and the infliction of reprisals. So much had the populations of occupied Europe suffered from the Axis forces' *excessive* recourse to these practices, that their representatives at Geneva would be satisfied with nothing less than their *total* prohibition. This may — far too little research has yet been done to allow even a sensible guess about what is obviously a very complicated matter — have done the standing of the law of war more harm than good. Likewise it is possible that the 1977 Additional Protocols, shaped under hot pressure of a different sort, have in some respects gone further than is militarily reasonable. The pressure this time came from outside Europe, from the 'national liberation movements' of the Third World, their ideological synpathisers in the First and Second, and the new states emerging from the ruins of the old empires and naturally rather critical of their former proprietors. The conferences leading up to 1977 thus became unprecedentedly politicised, and the interests or prejudices of that bloc determined some major issues in a sense unwelcome to the great NATO powers and also (what was not the same thing) to the ICRC itself. Armed conflicts against 'colonial domination and alien occupation and racist regimes', for example, were singled out as peculiarly just ones virtually of an international kind. This radically broke with the last two centuries' tradition of making no distinction between one sort of war and another; while non-international armed conflicts, far from receiving more consideration than in 1949, received rather less. In other respects, however, the law was proposed to be changed or clarified in ways more to the liking of the major military powers and solid old states; in the insistence, for example, that guerrillas wishing to be recognised as lawful combatants must observe the law themselves, and in the uncompromising ban on every sort of terrorism. It *was*, in the end, a compromise.

Whether the makers of the Additional Protocols went too far must be open to debate. The question, however, remains as valid about them as, through the years gone by, about the law of war and international humanitarian law in general. Of course they are imperfect. Given their circumstances and purposes, they cannot be anything but imperfect. But would mankind be better off without them? Would not an international arena stripped of the Red Cross and all that goes with it be a considerably worse world, more dangerous and inhumane, than the one we have now? The law of war, some powerfully argue, is normative rather than prescriptive. International humanitarian law, though it cannot yet presume to declare, can emphatically teach.

Notes

1. Jean Pictet, *The Principles of International Humanitarian Law* (International Committee of the Red Cross, Geneva, 1967), p. 13.
2. Ibid., pp. 14-15.
3. Ibid., pp. 19-24.
4. Jean Pictet, *The Fundamental Principles of the Red Cross* (Henry Dunant Institute, Geneva, 1979), p. 33.
5. Pictet, *Principles*, pp. 17-18.
6. Pictet, *Fundamental Principles*, p. 36.
7. Cited in Geoffrey Best, *Humanity in Warfare* (Weidenfeld and Nicolson, London, 1980), pp. 6-7.
8. David P. Forsythe, *Humanitarian Politics. The International Committee of the Red Cross* (Johns Hopkins University Press, Baltimore and London, 1977), p. 18.
9. Forsythe, *Humanitarian Politics*, p. 17, about Haiti; where, the author has been told, the President of the national society was found to be also head of Papa Doc's secret police.
10. See especially Background Papers 5 and 6, 'The Red Cross at National Level: a Profile', and 'As Others See Us: Views on the Red Cross'.
11. Pictet, *Fundamental Principles*, pp. 14-15.
12. Private information from the late Pierre Boissier of the Institut Henry-Dunant and the ICRC.
13. Milovan Djilas, *Conversations with Stalin* (Pelican, Harmondsworth, 1969), pp. 76, 87-8.
14. In the short story by 'Saki', 'Filboid Studge, the Story of a Mouse that Helped' in 'Saki' [H.H. Monro], *Complete Short Stories* (Bodley Head, London, 1930), p. 186.
15. Richard I. Miller, *The Law of War* (D.C. Heath, Lexington, Mass., 1975), p. 253; their names and qualifications are given on pp. 268-9. This problem, the apparent gulf between the premises of international humanitarian law and the cultural practices of different countries, was fleetingly glanced at in Geoffrey Best, 'Legal Restraints on Warfare: the Twentieth Century's Experience', *Journal of the Royal United Services Institute*, vol. 122, no. 3 (September 1977), pp. 3-9, 6-7. It was most regrettable that the footnotes were not printed. The book referred to there at the foot of p. 6 is, of course, Miller's.
16. Miller, *Law of War*, pp. 253-5.

17. Ibid., p. 256.
18. Ibid., pp. 259-60.
19. Ibid., pp. 264-5.
20. Ibid., p. 267.
21. *ICRC Bulletin*, no. 56 (3 September 1980), p. 1. Its President is cited more recently, in *ICRC Bulletin*, no. 66 (1 July 1981), as saying that 'international humanitarian law, which has been accepted as binding by virtually all States, has all too often been disregarded by its most fervent conference table champions'.
22. *ICRC Bulletin*, no. 45 (3 October 1979), reported that 21 Red Cross workers had been killed in action in the preceding eighteen months.
23. The author has attempted to tell some of it in his book mentioned in note 7 above.
24. The Additional Protocols may be found in British Command Paper 6927, Miscellaneous Papers no. 19 of 1977. They are summarised and commented on in Best, *Humanity in Warfare*, pp. 320-9; and by Peter Karsten, *Law, Soldiers and Combat* (Greenwood Press, Westport, Conn. 1978), pp. 146-67, 177-96.
25. Cited in Best, *Humanity in Warfare*, p. 296.

8 CONCLUDING COMMENTS

Ronald Hope-Jones

The reader who has conscientiously worked his way through these essays may well be feeling a little battered by now. If he has put at any rate a mental tick against propositions to which he assents, and a mental cross against those from which he dissents, he is likely to find that the ticks greatly outnumber the crosses, but that many of the propositions he has ticked are incompatible with one another. The aim of the present essay is to bring together the opposing arguments in those areas of debate in which the incompatible ticks are likely to be most numerous, and so assist the reader to decide where he really stands. But it is worth noting at the outset that the task would have been much harder if the various authors had not had the opportunity to discuss their original drafts with each other and with other members of the CCADD group concerned. Their revision of their drafts in the light of these discussions, and written comments, has done much to extend the area of consensus in the different essays and to isolate the issues on which fundamental disagreements still remain. Geoffrey Goodwin's essay, for example, had its origins in a background paper he wrote at an earlier stage in the exercise, when we were considering how 'The Search for Security' might be brought up to date. The final version owes not a little to comments made by members of the group, and in its present form most of us find we agree with most of it.

The Relevance of Christianity

These essays are written by Christians interested in the relationship between Christian principles, or at any rate practice, and problems of defence and disarmament; and many, if not most, of those who read them are likely to be Christians of similar interests. It therefore seems appropriate to start with Gordon Dunstan's essay, 'Theological Method in the Deterrence Debate', and with the first part of Bruce Kent's essay, 'A Christian Unilateralism'. The former, like Goodwin's essay, was originally written as a background paper at an earlier stage in our discussion, to help us consider whether there was, or indeed could be, a specifically 'Christian' approach to problems of security.

Concluding Comments 167

Dunstan begins by arguing that there are some human activities that cannot be discussed in Christian terms at all. War is an inherently unchristian pursuit, so that there can be no Christian way of prosecuting it; and more generally, what passes for a 'Christian' contribution to problem-solving is often no more than a veneer of Christian language upon policies dictated by other considerations. He then considers the claim that the Holy Scriptures provide a basis for a true 'Christian contribution', examining this claim first in relation to the Old Testament, then the New. His conclusion is that the political prescriptions of the Old Testament cannot be treated as divine commands of eternal validity, justifying holy wars, and that it is equally illegitimate to read off such prescriptions from the command to love or other words of Jesus. These 'are ingredients in a Christian ethics; but even a Christian ethics in its entirety cannot be transcribed directly, and without complement, into politics; and to quote such texts as though they could determine the issues of war and peace is to mistake the nature of Scripture, ethics and politics alike'.

Bruce Kent's standpoint is very different. 'Christian faith,' he writes, 'ought, nevertheless, to influence Christian judgement on "defence" issues as much as, and indeed sometimes more than, considerations of strategy, economics, psychology and military technology, important as those things are in any discussion about disarmament.' To be a Christian, he claims, 'is to become part of a living Body with its own norms and values in which the hidden Kingdom grows, inevitably, in conflict with the very different values of the world'. He then develops this theme in relation to two of these values, nationalism and violence, arguing that 'Christianity commits us . . . to the reality of internationalism as a priority over nationalism: to the priority of non-violence over violence'. But he is forced to admit that 'Unhappily, the vision of a Christian alternative set of values . . . is not one that has actually widely operated in practice . . . Christianity, wherever planted, has become a religion of conformity with the military and other policies of the nation-state.'

In the second part of his essay, dealing with the period since Christians became first tolerated, and then actively encouraged to take part in maintaining the fabric of the Empire, Dunstan argues that 'the "specifically Christian" contribution to the politics of security has been the application of Christian minds to the business of politics, *working within the terms and categories given them by politics itself* . . . ' It is not surprising that most members of our group, consisting as it did mainly of academics and senior civil servants, past or present, should have found themselves more in sympathy with Dunstan's views than with Kent's: a civil servant whose Christian convictions oblige him to challenge at every point the norms of the society he is trying to serve will soon have a nervous breakdown, if he is not sacked first. But that is not to say that all of us accept Dunstan's argument in its entirety. For example, at the beginning of his essay he writes, 'There is no specifically "Christian" way of waging war, or of amputating limbs, or of fixing oil prices or of deciding for or against the nuclear generation of energy.' But some ways of waging war are more open to Christian objection than others; Christians might be expected to condemn the amputation of limbs without anaesthetics (though it will be recalled that when anaesthetics were first used to relieve the pains of childbirth in the mid-nineteenth century, there was no lack of divines to condemn the practice as contrary to God's ordinance); a system of fixing oil prices that increases the burden on the poorest countries to the advantage of the richest does not have much to commend it from a Christian point of view; and a Christian's belief, as Kent puts it, that we are 'short-term stewards, partners with the Lord in his ongoing work of creation', cannot but influence his views on the nuclear generation of energy, bearing in mind the pollution problems involved. To the objection that the word 'Christian' can be written out of the above sentence by substituting 'humanitarian' or 'ecologist', as appropriate, the Christian may reply that his own ethical beliefs derive from his Christian faith, even if other people arrive at very similar beliefs from a different standpoint. But once he admits that non-Christians share his ethical approach to the four activities which Dunstan takes as examples, he surely has to accept Dunstan's argument that there is no *specifically* Christian way of conducting them.

The same basic issue is raised in Geoffrey Best's essay, when he discusses Jean Pictet's views on the principles of humanitarian law. Best takes Pictet's glad acceptance of Max Huber's analogy of the Good Samaritan as evidence of the Christian sources of his presentation of these principles, but points out that Pictet takes care to avoid imputing to humanitarianism anything like a primarily Christian character. Most of us would regard 'Whatsoever ye would that men should do to you, do ye even so to them' as a specifically Christian precept; but as Pictet says, it is a precept which 'can be found, in almost identical form, in all the great religions . . . It is also the golden rule of the positivists . . . ' This is precisely the point that Gordon Dunstan makes with regard to a wider field of human activity.

Another possible objection to Dunstan's argument is that he dismisses too lightly the relevance of the Old Testament. He writes,

> The prescriptions for security were *political* prescriptions: invade this territory, respect the frontiers of another; go out to battle, refrain from battle; ally yourself with this nation, do not become entangled with another . . . That was the hard political advice, given out of political and military experience. It was given also out of religious conviction, expressed in unremitting calls for *faithfulness*, an active fidelity to YAHWEH, their own God . . .

Dunstan is clearly right in insisting that the tribal bellicosities and political prescriptions of the Old Testament must not be treated as divine commands. Both the politico/military situation and our perceptions of God's will have altered since the days of Isaiah and Jeremiah. But is it inconsistent with exegetical and theological integrity to hold that one of the lessons to be learnt from the Old Testament is that religious conviction is a legitimate (though obviously not the sole) determinant of defence policy?

Two further rather obvious points should perhaps be made to help the reader make up his mind where he himself stands in this debate. First, it is not a question of having to make a straight choice between Dunstan's position and Kent's. There are many other possible positions, some intermediate, others more extreme in either direction. Second, to accept the theological views of either is not necessarily to accept their political prescriptions, explicit or implicit. For example, Kent's unilateralism does not derive solely, or even perhaps primarily, from his religious conviction, but from his judgement that the arms race is likely to lead to a nuclear holocaust, and that the prospects for multilateral disarmament are not such as to justify a reliance on that alone to save us.

The East-West Conflict

All the essays in this volume, except Geoffrey Best's, are concerned, in one way or another, with deterrence, and in particular with nuclear deterrence, as a means of preventing war between East and West. But there are two classes of Christians who may think that deterrence is quite unnecessary. First, there are those who regard both Western and Eastern institutions and societies as falling so far short of the ideal Christian society which reflects the values of the Kingdom of God that there is nothing to choose between them; in which case neither has anything that is worth defending against the other, and there is therefore no moral justification for mutual deterrence. Second, there are those who think that the Soviet military build-up, even if it is a reality and not a mere invention of Western military intelligence, represents no more than a natural reaction to what is perceived as a Western military threat; and that the Russians are as anxious as we are to avoid an East-West war and have no intention whatever of initiating one by invading Western Europe; so that again there is no need for deterrence.

Traces of both attitudes are detectable in Kent's paper, particularly in the second section where he writes about 'Us and Them'. He does well to remind us of the defects of our own societies and political systems, since there can be no religious justification for defending them if they are indeed irredeemably corrupt. But he sometimes writes as if there was nothing to choose between Western democracy and Soviet totalitarianism. This seems an odd judgement for any Christian to make, let alone one who must surely recognise that anyone who tried to mount a Campaign for Nuclear Disarmament in the Soviet Union would be lucky if he only found himself in a psychiatric hospital. It has not been the purpose of these essays to convince anyone that Western democracy is preferable to Soviet totalitarianism, and I shall not pursue the matter further here. One reason for not doing so is that our readers are unlikely to include many who doubt it. But there may well be a larger number who think that the Soviet military build-up is to be explained in terms of the West's aggressive posture, and nothing more.

This is a question, first, of capability and, second, of intention. As regards capability, the evidence for the Soviet build-up comes from neutral international sources, such as the Stockholm International Peace Research Institute, and is not denied by the Soviet government itself. As regards intention, no one supposes that the Soviet leaders want a nuclear war with the West. But in the absence of a convincing Western deterrent, the Soviet government could exercise irresistible political pressure, backed by the nuclear threat; and in the light of its Marxist-Leninist convictions and professed aims, it is prudent to suppose that it might seek to do so. And even if the present Soviet government's intentions are entirely pacific, who can say whether this will still be the case in two years' time, let alone ten or twenty? Of course, the argument can be reversed. It is an article of faith in the Marxist-Leninist creed that Western capitalism is dedicated to the destruction of Soviet Communism, so that in the absence of a convincing *Eastern* deterrent the *West* could exercise irresistible political pressure, etc. Nor is this view based only on ideological preconceptions, as is clear from the section in Goodwin's essay entitled 'Soviet Perceptions', which should be re-read by anyone who thinks that these essays present only a Western point of view. The only conclusion the reader can draw, I suggest, is that, given the present atmosphere of profound mistrust, rooted partly in nationalism and partly in ideology, between East and West, it is regrettably inevitable that both sides should think it necessary to equip themselves with an adequate deterrent. But though it is quite understandable that neither the United States nor the Soviet Union should wish to be significantly inferior to the other in military terms, it is much less easy to understand or justify a desire on the part of either to be significantly superior.

Since the above observations reflect the main area of disagreement between Bruce Kent and the rest of the group, it may be appropriate to end this section by noting that we agree entirely with his condemnation of what he calls 'crusading fervour'. In the section of his essay entitled 'The Ideological Dimension' Geoffrey Goodwin notes with regret the resurgence of an earlier tendency to regard the East-West conflict as a conflict of good with evil. Again, Gordon Dunstan writes, 'This realism about man is a preservative, also, against false polarities, dividing men and nations into angels and devils, sons of light and sons of darkness, with ourselves wholly good, wholly right, and others, the enemy or potential enemy, wholly bad, wholly wrong.' This is very much the point that Kent himself makes.

The Ethics of Deterrence

The object of deterrence is to prevent war — the word itself means 'frightening off' — but it is inherent in the concept of deterrence that, if the opponent is not frightened off, the weapons that constitute the deterrent will be used. The hope, always, is that actual use will not be necessary; but if the deterrent is to be effective, use must always be seen as a possibility, not to be discounted, by the opponent. In considering the ethical issues involved in nuclear deterrence, we may look first at those involved in the actual use of nuclear weapons, then at those involved in the mere possession of such weapons, with particular reference to the assumption that possession implies a conditional intention to use them.

What moral judgement, then, is to be passed on the state, or alliance, that has a nuclear capability, that makes clear its intention to use that capability against an aggressor, and that then uses it when attacked with nuclear weapons? At one end of the scale, we have the argument that moral responsibility rests wholly with the original aggressor. In the first draft of his essay, Gordon Dunstan wrote,

> The *intention* is, by maintaining a credible threat, to prevent any occasion for its use — to deter the other side from the first, immoral, act, the nuclear strike. If, knowing the consequences, he commits that act, the responsibility for the consequences is his; he brings undiscriminating destruction upon his own head, and on all his people. While he is deterred from evil — a nuclear attack — by the threat of nuclear reaction, the maintenance of the threat could be morally justified. To deny justification for carrying out that threat would be to rob the deterrent of its force . . .

At the other end of the scale, we have Barrie Paskins's position, which is that the reaction of the possessor of a nuclear deterrent when subjected to a first nuclear strike is not an automatic one; the decision whether or not to retaliate, to implement the conditional intention, is still open; and the moral responsibility for taking that decision, and consequently for the undiscriminating destruction caused if the decision is to make a retaliatory nuclear strike, rests wholly with the possessor of the (failed) nuclear deterrent. I have quoted Dunstan's original version of his essay; but in subsequent versions he changed 'the responsibility for the consequence is his' to 'the responsibility for the consequence is primarily his, however much the respondent is also to blame'. Most members of our group would probably opt for some such intermediate position, though we would not necessarily use the word 'primarily'. The reader must decide for himself where he stands in this debate. (God's answer, alas, is not available: but He presumably hopes, as we do, that He will not be required to give one.)

It will be noted that in the passage quoted above Dunstan wrote 'The intention is . . . to deter the other side from the first, immoral, act, *the nuclear strike*.' But what are we to make of the fact that the NATO allies have steadfastly refused to give any undertaking that they will not be the first to use nuclear weapons? The reason for their refusal is plain enough. Whatever the exact figures may be, it is an uncontrovertible fact that the Warsaw Pact countries have a marked superiority over the NATO allies both in numbers and in conventional weapons, particularly tanks. Were they to launch a full-scale invasion of Western Europe with conventional forces, it might be only a matter of

Concluding Comments

days before a decision had to be taken whether to cross the nuclear Rubicon, and use battlefield nuclear weapons. Would this be morally justified?

In his address to the national convention of the World Disarmament Campaign in April 1980, Cardinal Hume argued that some uses of nuclear weapons would not involve the destruction of civilian populations, and might therefore be justified without too great moral difficulty. It was possible to argue, he admitted, that the discriminate non-strategic use of nuclear weapons might end conflict by persuading the aggressor to desist in order to avoid unacceptable harm — the 'intrawar deterrence' theory, as explained in Hockaday's and Paskins's essays. But though, on these grounds, he refused to condemn outright the possession of nuclear weapons directed to military targets, he laid down two vital conditions: 'First, their possession and use would not be justified unless it is possible in practice to draw a clear distinction between military installations and personnel who will be destroyed and the civilian population which may be affected. And secondly, the use of strategic weapons of this type must not lead to escalation.' But as regards the first of these conditions, it is generally accepted that there is little prospect of even tactical nuclear weapons being used in Europe without collateral damage to the civilian population; while as regards the second (and it is interesting, incidentally, that Cardinal Hume appears to regard the first user of nuclear weapons as morally responsible for his opponent's reaction), there can surely be no circumstances in which the first user can be certain that there will be no escalation; indeed, the danger of escalation is essential to the deterrent effect of the weapons. It is precisely because of our doubts about the possibility of Cardinal Hume's two conditions being met that several of us, who accept the need for a NATO nuclear deterrent, nevertheless find it distressing that the effectiveness of this deterrent should rest on NATO's willingness to make first use of nuclear weapons. Others find it equally distressing that, as they see it, the effectiveness of NATO's overall deterrence of aggression is weakened to the extent that it rests on this basis.

We come now to the moral issues involved in the mere possession of nuclear weapons of mass destruction. Here some would argue that, provided the sole object of having such weapons is to prevent aggression, it cannot in itself be described as wicked, however wicked their actual use might be, even in retaliation. This is not Barrie Paskins's view. He writes,

Normally, one thinks that if an action is immoral then the intention to commit that action is also immoral. Not only highway robbery but also the robber caught before he can effect his intention is rightly punishable. Do the laudable aim and well-intentioned complexity or strength of deterrence soften the adverse judgement one would normally make? I am unable to believe that they do . . .

But does the possession of a nuclear deterrent necessarily involve a conditional intention to use nuclear weapons? Paskins is satisfied that it does, and dismisses the possibility of bluff out of hand. 'This commitment to nuclear weapons', he writes, 'is not a bluff: political will is tested remorselessly and any bluff will be called.' But how can my opponent know that I am bluffing unless he is prepared to take the risk that I am not? For the sake of simplicity, let us take the British independent strategic nuclear deterrent as our example. If it is indeed the case, as is generally supposed, that only the Prime Minister can set in motion the train of events that leads to the launching of our Polaris missiles, it is quite conceivable, in view of the appalling scale of the retaliatory nuclear assault that would inevitably follow, that the Prime Minister at any given time may be unable to envisage any set of conditions in which he or she would give the order to fire, or may indeed be absolutely determined not to give such an order in any circumstances. It can be argued that this need not degrade the efficacy of the deterrent, since this depends entirely on the perceived capability, and not on any real conditional intention to exercise that capability. 'Worst-case analysis' should lead the aggressor to suppose that the Prime Minister is not bluffing, even if in fact he is.

Paskins will have none of this. He argues that in public our leaders insist on their resolve to use nuclear weapons if necessary, that careful thought must have been given to the circumstances in which it will be 'necessary' to use them, and that 'taken together, the public pronouncements which are there for all to see and the secret positions which must exist amount to a conditional intention.' But the public pronouncements do not have to be taken at their face value; they may equally well be taken as a necessary part of the bluff, to encourage the opponent to believe in a conditional intention that does not in fact exist. And as regards his 'secret positions', I am as entitled to assume that there are no carefully prepared plans for national suicide as he is to assume that there are.

Whether or not the above argument, based on the negative intention of the person responsible for the final decision, is considered valid, it does not, as Paskins points out, satisfactorily dispose of the moral

Concluding Comments

objection, since whatever the Prime Minister's intention may be, there are hundreds of other people, including the crews of the submarines, who cannot be certain that the order to fire will never be given and who must be assumed to have a conditional intention to play their part in executing the order, if it is given. And in executing the order they would be acting on our behalf. Thus all those who acquiesce in the decision to maintain our independent nuclear deterrent have some share in whatever wickedness may attach to its possession — with the possible exception of the Prime Minister. And all those, a much larger number, who accept the need for the NATO nuclear deterrent must similarly accept their share in whatever wickedness may attach to that. There is no moral kudos to be gained from scrapping our own nuclear weapons and relying entirely on those of the Americans.

Most of the members of the group accept the distinction which Hockaday, following W.D. Ross, makes between the *right* and the *good*, and would agree with him that 'although the conditional intention may contain an element of moral evil, a strategy of deterrence involving the conditional intention may be the most effective way of securing the twin objectives of preventing war and checking political aggression and may therefore be a morally acceptable price to pay to achieve those objectives'. Where those of us who take this line would disagree among ourselves is in our assessment of the degree of moral evil attaching to the possession and conditional intention to use nuclear weapons. One suggestion made in our discussions was that this varies with the probability we attach to the weapons actually being used. If it is thought that the probability of the Soviet Union invading Western Europe or applying unacceptable political pressure in the absence of an effective deterrent is two-thirds, and that with the deterrent the probability of such action being taken *and* of the West's response leading to a nuclear holocaust is only one in a hundred, there would seem to be greater moral justification for the deterrent than if it is thought that these probabilities are, say, one in five and one in ten respectively.

Multilateral and Unilateral Disarmament

Whatever other disagreements there may have been between the members of our group, we would all endorse the following propositions:

(i) that the nuclear arms race between East and West is wasteful both

of material and of human resources, and therefore immoral;
(ii) that it is potentially destabilising and therefore dangerous;
(iii) that every attempt should be made to achieve a stable balance of deterrence at a much lower level.

These propositions may seem so self-evident as to be not worth stating, but we should remember that not a few Christians would contest (iii), arguing that deterrence based on the conditional intention to use weapons of mass destruction is inherently sinful, and that it is accordingly our moral duty, regardless of possible consequences, to disssociate ourselves entirely from the nuclear deterrent, leaving NATO if our allies cannot be persuaded to follow a similar course. But even those who take this extreme line will recognise that at the moment they are in a minority, and that in a democratic state the government must seek to provide security against aggression for its citizens, if that is what most of them want. It is difficult to see how that security could be maintained if NATO got rid of its nuclear arsenal and the Warsaw Pact countries retained theirs. These three propositions may therefore be taken as establishing a basis of common ground which most multilateralists and unilateralists would accept.

Bruce Kent argues that it is wrong to regard the two doctrines as being in stark opposition, since the unilateralist is in favour of multilateral disarmament, if it can be achieved. But the essence of the unilateralist case is that it is unsafe to rely only on multilateral negotiations, since the experience of the last thirty-five years gives us no reason to suppose that multilateral negotiations will succeed in the future in reversing the technology-led nuclear arms race which it has conspicuously failed even to check in the past. The essence of the multilateralist case is that any significant measure of unilateral nuclear disarmament by the West would endanger the deterrent balance, and could result in the Soviet Union achieving a first-strike capability, that is the ability, by striking first, to eliminate so much of the West's nuclear arsenal that its retaliatory strike would no longer cause unacceptable damage.

At the heart of this debate lies the question whether the balance of deterrence should be regarded as delicate or robust. Roy Dean, a multilateralist, regards it as delicate; while Barrie Paskins, a unilateralist, regards it as robust. But it would be wrong to assume that there are two simple equations, multilateral equals delicate, and unilateralist equals robust. There are some who regard the balance of deterrence as so inherently and dangerously delicate and unstable that they deny the possibility of its providing any meaningful security and so advocate complete unilateral nuclear disarmament by the West, on 'better red than dead' grounds. Equally there are others, including some who have studied the theory of deterrence in greatest depth, who regard the balance as robust, but are nevertheless convinced multilateralists. They would argue that the deterrent balance is robust, in the sense that there is no likelihood of a technical breakthrough of such dimensions as to give either East or West a credible first-strike capability, but that the objective must be a progressive scaling-down of the nuclear forces on both sides; that this can only be achieved by a series of negotiated agreements for mutual reductions, starting from a position of strategic equivalence if not strict numerical parity; and that if the NATO allies were to make significant unilateral cuts, they would deprive themselves of the leverage necessary to induce the Warsaw Pact countries to match these cuts, and quite possibly leave themselves in a position in which the Russians could derive at any rate political advantage from the military superiority gratuitously given them.

There is, of course, a great deal more than this to be said about the delicacy/robustness controversy. Much of Paskins's essay is concerned with this question, either directly or indirectly, and it is impossible to summarise his detailed arguments here. But there are two points that may be made. First, the credibility of the British independent nuclear deterrent seems to imply an extremely robust view of the deterrent balance. At any one time we may have only one of our Polaris missile submarines at sea. The disparity between our strategic nuclear capability and that of the Russians is enormous, yet it is claimed that ours constitutes a credible deterrent. If the claim is justified, why should NATO require such a vastly larger strategic nuclear deterrent, even allowing for the fact that NATO's deterrent is directed towards a much wider range of possible aggressions? Second, if the deterrent balance is seen as robust, as it must be if unilateral disarmament initiatives are advocated, how can the unilateralist argue that the continuation of the arms race is likely to end in nuclear conflagration? He can hold such a view only if he thinks the balance is delicate, and so capable of being destroyed by a technological breakthrough.

I suspect that at any rate part of the answer to both these points may lie in the fact that the terms 'robust' and 'delicate' appear to be used in different senses at different times. The balance of deterrence may be described as delicate by those who think that one side or the other may at any time acquire a first-strike capability, and as robust by those who think this unlikely. But it may equally be described as delicate by those who think there is a high probability of one side making a mistaken assessment of the other's willingness to use nuclear weapons in a crisis, and as robust by those who think there is only a remote possibility of this. There is no illogicality in regarding deterrence as robust in the first sense, but delicate in the second.

One point on which both multilateralist and unilateralist must surely agree is that as the level of the deterrent balance is progressively lowered, it becomes more delicate; the risk of disaster increases as the reduction of the nuclear capability on both sides approaches the point at which a sudden increase in the capability of either would give it a first-strike capability, or be thought by its opponent to do so. The reasons for this and its consequences for disarmament by stages are explained by Arthur Hockaday in the section of his essay in which he draws on the work of Legault and Lindsey, and reproduces two of their diagrams. While these diagrams should certainly help the reader to understand the theory, a word of warning is perhaps called for. No one should imagine that the Americans and Russians could now sit down at the same table, with identical diagrams in front of them, mark in the point represented by the numbers of their respective missiles and proceed to chart a safe course to a minimum mutual deterrent. The balance of mutual deterrence is not just a question of relative numbers of missiles, or even warheads of a known accuracy and destructive power. It rests also on a number of psychological imponderables, quantifiable only by the crude guesswork of professional pessimists.

How then are we to proceed? Some unilateralists will argue that the West should take the initiative by making a significant reduction in its nuclear forces, but should make no further reduction until there is an adequate response from the East. Others, pressing the robust view of deterrence to the extreme, would urge that NATO should go further than this and reduce its nuclear capability, either gradually or at one stroke, to something approaching a minimum nuclear deterrent and not worry if there is no Soviet response. The multilateralist will have none of this. The lesson that Roy Dean would have us learn from his factual review of the arms control negotiations since the Second World War is that any unilateral reduction in its nuclear forces that the West made would not be matched by the Russians, and that the greater the unilateral reduction made, the greater the danger that they would react by trying to exploit our self-imposed weakness, regarding this as an entirely appropriate and historically determined rectification in the international correlation of forces – and thus one of the positive changes in world politics that Marxism/Leninism foretells.

Concluding Comments 179

One point worth making at this stage is that the multilateral/unilateral debate is all too often bedevilled by imprecision in the use of the word unilateralism. Multilateralists constitute a relatively homogeneous group, but unilateralists come in all shapes and sizes, ranging from the two classes referred to in the preceding paragraph to those who think that NATO should scrap all its nuclear weapons overnight, and that failing this Britain should pull out of NATO. The issue is further confused by the fact that those who are opposed to the Trident programme for whatever reason, are frequently referred to as unilateralists. It needs to be made clear that there are many, including some members of our group, who think that the government's decision to embark on this programme is a mistaken one (for example because the only object of the British independent nuclear deterrent is to ensure that we are forced to surrender by starvation or conventional warfare, rather than by nuclear assault, if the NATO deterrent fails and the continent of Europe is overrun), but who are nevertheless firm multilateralists in the NATO context. There is no illogicality in such a position, since there is no *quid pro quo* to be extracted from the Russians by negotiation.

The most enthusiastic advocates of nuclear disarmament are all too often those who know least about the attempts that have actually been made over the past thirty years to negotiate worthwhile arms control agreements. Such readers should find Roy Dean's essay particularly valuable, and if they feel tempted, at the end of it, to make disparaging remarks about mountains and mice, let them bear in mind a sentence in Paskins's essay to the effect that 'those who consider the achievements of arms control to date meagre tend to underestimate the sheer difficulty of the thing'. Elizabeth Young (to whom all our essayists are indebted in greater or lesser degree for her perceptive comments) has suggested to us that one reason for the comparative paucity of results is that the negotiating machinery is defective. She has been greatly impressed by the large measure of success achieved, after years of patient labour, by the UN Commission on the Law of the Sea, and thinks that if the arms control negotiations were dealt with on similar lines, by a Committee of the Whole and appropriate subcommittees, using the negotiating technique of a 'single, informal, negotiating text', this would allow greater pressure to be exerted on the United States and the Soviet Union, without whose agreement no progress can be made. Others may doubt whether any structural change in the negotiating machinery would cause the two superpowers to make any concessions to the majority view which they would not otherwise make, in a field in which their vital security interests are so closely involved.

However that may be, Dean himself admits that the immediate prospects for arms control are not bright. His argument is not so much that multilateral negotiations are bound to end in success, if we try hard enough and long enough, as that any quick results apparently achieved by unilateral disarmament measures will prove illusory, and will take us towards Armageddon rather than away from it. The opposite view is that multilateral negotiations are unlikely to be more successful in the future than they have been in the past, that the time has come for the West to try the unilateral approach, and that this can be done without any real sacrifice of security. It is for the reader to make his own judgement. As far as our group is concerned, I think it is true to say that some who started out as staunch multilateralists became at any rate more sympathetic towards the less extreme forms of unilateralism, as our discussions proceeded.

Most of this section has been concerned with questions of prudence rather than principle. But at the beginning of it I referred to Christians who maintain that deterrence based on the conditional intention to use weapons of mass destruction is inherently sinful and that it is accordingly our moral duty, regardless of the possible consequences, to dissociate ourselves entirely from the nuclear deterrent. Paskins regards nuclear deterrence as inherently sinful, but he does not draw such a conclusion, explaining his refusal to do so by his analogy of the man conducting two adulterous affairs. However, he offers no suggestions on how we can ever move from a position of stable mutual deterrence, based on the conditional intention to use weapons of mass destruction, to one in which such weapons have actually been eliminated. That being so, it seems fair to ask what right he has to criticise multilateralists on the grounds that ' "For the foreseeable future" they expect nuclear weapons, and the conditional intention to use nuclear weapons, to remain a vital part of their strategy'; or to complain at their 'implicit readiness to continue with this immoral conditional intention for an unlimited future'. To put the criticism another way, the theme of Paskins's essay is accurately described by its title 'Deep Cuts are Morally Imperative'. But deep cuts still leave the party that makes them with weapons of mass destruction that it has the conditional intention to use. Admittedly there are fewer of them, but the inherently sinful conditional intention is quite unaffected by the cuts.

Some sort of an answer to such criticisms could perhaps be made on the following lines. The most we can hope for in the next ten or twenty years is a continuation of stable mutual deterrence at a progressively lower level. We shall only be able to move on to the elimination of the nuclear deterrent when the present climate of mutual suspicion has been replaced by one of mutual trust. The acceptance by the West of a position of clear nuclear inferiority, initiated by a substantial measure of unilateral nuclear disarmament (within the limits of Legault/ Lindsey's area of stable mutual deterrence), offers the best chance of convincing the Soviet government that our aims are not in fact aggressive. It would thus do more than any other step we might take to allay Soviet fears and establish the climate of trust which is a precondition for the eventual elimination of nuclear weapons. Several members of our group accept the thesis that the Christian should feel a particular responsibility for establishing and strengthening the 'sinews of trust', and that he should be prepared to take greater risks for peace — real peace, not the sort based on nuclear deterrence — than the non-Christian. They, at any rate, will not lightly dismiss Paskins's claim that 'deep cuts are a moral imperative'.

Wider Perspectives

As Geoffrey Goodwin explained in the introduction, the present volume of essays originated in an abortive attempt to produce a revised version of a British Council of Churches report of 1972 entitled *The Search for Security*. One of the reasons why this attempt proved abortive was that, try as we might in our discussions to widen our perspective, we always seemed to come back to the theory, practice and ethics of deterrence and to the multilateralist/unilateralist debate. In short, we were preoccupied, as most of our readers are likely to be, with the nuclear dimension of the East-West conflict.

Even from the limited point of view of the East-West conflict, this obsessive concentration on the nuclear dimension needs to be put in perspective. Many people seem to think that if only nuclear weapons could be brought under control, all would be well. But we have to remember that nuclear weapons are only the most spectacularly horrific items in the NATO and Warsaw Pact arsenals. It is arguable that even if they had never been invented, relations between East and West would be much the same as they are today, with each side relying for its security on a massive deterrent array of battlefield, theatre and strategic missiles, armed with warheads containing chemical or biological agents. Goodwin and Paskins both make the point that the nuclear race itself foments the mutual distrust of which it is the product; but any other sort of arms race would have much the same effect. Peace will only dawn when neither side has any reason to fear the other's aggressive intentions.

No one supposes that the East-West conflict is just a matter of two heavily armed alliances eyeing each other with mutual suspicion across the Iron Curtain. It is a conflict that reaches out into every corner of the globe. Goodwin, in the section of his essay entitled 'Order and Tension in International Society', draws attention to the way in which superpower rivalry has made more difficult the orderly management of change in the Third World, and the reader may wish to look again at the passage in Dunstan's essay on the impact of the advent of nuclear weapons on the traditional 'just war' condition of proportion, in relation to Third World countries. He writes

> The harm likely to follow, in an era of nuclear weapons — and even of highly destructive 'conventional' weapons — from armed intervention to repel the invader would be so disproportionate to any foreseeable good that the invaded and oppressed must be left to their fate. Sustained protest, and the mildest of cultural and economic sanctions, are the most that such an aggressor has now to fear — unless he threatens one of the Western powers directly.

This reluctance to intervene may be one of the reasons why the tide of violence in the Third World has risen rather than fallen since the Second World War. It is to this area of actual conflict, rather than hypothetical holocaust, that Geoffrey Best's essay, which to the casual reader may seem out of place in the present volume, most usefully directs our attention. He points out that in the last thirty-five years between 10 and 20 million people, most of them civilians rather than soldiers, have died in what we in the West are apt to regard as 'local wars'. It is a statistic that will come as a shock to many. His concern is not with the causes of such wars, or the moral justification for them, or even with how they might be prevented, but quite simply with how the human suffering involved in them can be minimised. Most people, if asked for their views on the law of war and attempts to strengthen and enforce it, would unthinkingly reply that in the nuclear age it had become irrelevant and futile. It is a sad reflection of this general attitude that when a UN Convention (based on an Anglo-Dutch draft) was signed in April 1981, providing for an extension of the restrictions on the use of some of the more unpleasant items in the 'conventional' armoury, *The Times* did not even think it worth reporting. No one

Concluding Comments

who reads Best's essay is likely, in future, to discount the value of even such a modest measure as the 1981 Convention, let alone the ICRC's heroic efforts over the years to reduce the suffering involved in wars of every kind. Equally, however, no one is likely to underestimate the difficulty of making the various Geneva Conventions fully effective, when these are in conflict with popular attitudes rooted in traditional culture. Best argues that the law of war is a delicate plant, even when nurtured in its natural habitat, and that its frequent failure to take root when transplanted into an unfavourable environment is hardly surprising. However, he also quotes, with apparent approval, Jean Pictet's view that the humanitarianism which underlies the Geneva Conventions is not specifically Christian, but common to all the major religions. If this is indeed the case, should not all these religions be co-operating in the task of trying to bring about increased popular understanding of and support for the law of war accepted by national governments? And should not Christian churches be giving the lead in this?

Best draws a distinction between the 'Geneva man', concerned only with the relief of suffering, and the 'Hague man', who seeks the 'awkward compromise between humanitarian principle and military reality' and who is obliged to wrestle with the paradox of restraining violence in a situation which by definition legitimises it. The Christian paradigm of the Geneva man is the Good Samaritan, but to the Hague man, Best writes, 'The Good Samaritan has nothing useful to say . . . neither have the Decalogue or the Beatitudes. He dare recall them only through a venerable filter of casuistical or dialectical interpretation.' It is clear that in so far as we who have written these essays, or contributed to them by our discussions, are all concerned with the grim realities of international politics in a fallen world which spends 350 billion dollars a year on maintaining a capability to make war, we are more akin to Hague than to Geneva man. Are we then guilty of casuistical or dialectical interpretation when we try to look at problems of defence and disarmament in the light of our Christian beliefs? We do not think so. The Good Samaritan may not be of much assistance to us in our search for solutions to these problems, but there are other strands in the Gospel narratives that must not be forgotten. As Best himself reminds us, when Jesus healed the centurion's servant (Luke 7:1-10), He commended him for his faith; He did not tell him to resign his commission. If the Good Samaritan is the Christian paradigm of Geneva man, then the Good Centurion, bridging the gulf between Caesar's world and God's world, may be taken as the paradigm of Hague man. We may perhaps derive some comfort from the fact that for the past eighty years the two streams of humanitarian international law — the Geneva stream and the Hague stream — have been converging.

APPENDIX: EXCERPTS FROM SOME OF THE CHURCH STATEMENTS REFERRED TO BY CONTRIBUTORS

Resolution of the Lambeth Conference, 1968

This Conference

(a) reaffirms the words of the Conference of 1930 that 'war as a method of settling international disputes is incompatible with the teaching and example of Our Lord Jesus Christ'.
(b) states emphatically that it condemns the use of nuclear and bacteriological weapons.
(c) holds that it is the concern of the Church
  (i) to uphold and extend the right of conscientious objection.
  (ii) to oppose persistently the claim that total war or the use of weapons however ruthless or indiscriminate can be justified by results.
(d) urges upon Christians the duty to support international action either through the United Nations or otherwise to settle disputes justly without recourse to war; to work towards the abolition of the competitive supply of armaments; and to develop adequate machinery for the keeping of a just and permanent peace.

Extract from Encyclical Letter of Pope John XXIII on Human Rights and Duties *Pacem in Terris* (1963)[1]

41. The result is that people are living in perpetual dread of a storm bursting upon them at any moment with frightening fury. No wonder, since the deadly weapons are there already. If it is hard to believe that there are men prepared to take the responsibility for the slaughter and devastation that war would bring, we cannot get away from the fact that the fires of war could be sparked off by some incident arising from a misunderstanding. Besides, however much the stupendously destructive potential of modern armaments may at the moment act as a deterrent to war, there is reason to fear that, unless there is an end to the testing of nuclear weapons, various kinds of life on earth will be imperilled . . .

112. Hence justice, right reason and an appreciation of human

dignity all clamour for a stop to be put to the competition in military strength; for a simultaneous reduction by all states in the stockpile of arms; for a ban on atomic weapons and for agreement finally on a suitable form of universal disarmament mutually and effectively monitored. 'We must work with might and main,' said our predecessor, Pius XII, 'to prevent the human family from being devastated by a third world war that will bring immeasurable disruption in social and economic affairs, countless outrages and moral havoc.'

Extract from *Gaudium et Spes* (1965)[2]

Total War

80. The horror and perversity of war are immensely magnified by the multiplication of scientific weapons. For acts of war involving these weapons can inflict massive and indiscriminate destruction far exceeding the bounds of legitimate defense. Indeed, if the kind of instruments which can now be found in the armories of the great nations were to be employed to their fullest, an almost total and altogether reciprocal slaughter of each side by the other would follow, not to mention the widespread devastation which would take place in the world and the deadly after-effects which would be spawned by the use of such weapons.

All these considerations compel us to undertake an evaluation of war with an entirely new attitude. The men of our time must realize that they will have to give a somber reckoning for their deeds of war. For the course of the future will depend largely on the decisions they make today.

With these truths in mind, this most holy Synod makes its own the condemnations of total war already pronounced by recent Popes, and issues the following declaration:

Any act of war aimed indiscriminately at the destruction of entire cities or of extensive areas along with their population is a crime against God and man himself. It merits unequivocal and unhesitating condemnation.

The unique hazard of modern warfare consists in this: it provides those who possess modern scientific weapons with a kind of occasion for perpetrating just such abominations. Moreover, through a certain inexorable chain of events, it can urge men on to the most atrocious decisions. That such in fact may never happen in the future, the bishops of the whole world, in unity assembled, beg all men, especially govern-

Appendix

ment officials and military leaders, to give unremitting thought to the awesome responsibility which is theirs before God and the entire human race.

The Arms Race

81. Scientific weapons, to be sure, are not amassed solely for use in war. The defensive strength of any nation is considered to be dependent upon its capacity for immediate retaliation against an adversary. Hence this accumulation of arms, which increases each year, also serves, in a way heretofore unknown, as a deterrent to possible enemy attack. Many regard this state of affairs as the most effective way by which peace of a sort can be maintained between nations at the present time.

Whatever be the case with this method of deterrence, men should be convinced that the arms race in which so many countries are engaged is not a safe way to preserve a steady peace. Nor is the so-called balance resulting from this race a sure and authentic peace. Rather than being eliminated thereby, the causes of war threaten to grow gradually stronger.

While extravagant sums are being spent for the furnishing of ever new weapons, an adequate remedy cannot be provided for the multiple miseries afflicting the whole modern world.

Extract from Pope John Paul II's World Peace Day Message for 1979

Statesmen, leaders of peoples and of international organisations, I express to you my heartfelt esteem, and I offer my entire support for your often wearisome efforts to maintain or re-establish peace. Furthermore, being aware that mankind's happiness and even survival is at stake, and convinced of my grave responsibility to echo Christ's momentous appeal, 'Blessed are the peacemakers,' I dare to encourage you to go further. Open up new doors to peace. Do everything in your power to make the way of dialogue prevail over that of force. Let this find its first application at the inward level: how can the peoples truly foster international peace, if they themselves are prisoners of ideologies according to which justice and peace are obtained only by reducing to impotence those who, before any examination, are judged unfit to build their own destinies or incapable of co-operating for the common good? Be convinced that honour and effectiveness in negotiating with opponents are not measured by the degree of inflexibility in defending one's interest, but by the participants' capacity for respect, truth, benevolence and brotherhood – or, let us say, by their humanity. Make gestures of peace, even audacious ones, to break free from vicious circles and from the dead weight of passions inherited from history. Then patiently weave the political, economic and cultural fabric of peace. Create – the hour is ripe and time presses – ever wider areas of disarmament. Have the courage to re-examine in depth the disquieting question of the arms trade. Learn to detect latent conflicts in time and settle them calmly before they arouse passions. Give appropriate institutional frameworks to regional groups and the world community. Renounce the utilisation of . . . spiritual values at the service of conflicts of interests, values which are then brought down to the level of those conflicts and make them more unyielding. Take care that the legitimate desire to communicate ideas is not exercised through the pressure of threats and arms.

Extract from Cardinal Hume's Address to the World Disarmament Campaign 12 April 1980[3]

But the Christian realist cannot simply condemn but must reflect on those relevant and urgent issues of war and peace. I do not think that I am a pacifist. I do not think one can reject categorically and without qualification the right to self defence. I believe an individual retains this right personally and I believe the state has it on behalf of the citizens. In fact I want to express admiration and gratitude to all those men and women who fought and died for the freedom of our country in two world wars. We too easily forget that we live in freedom today because of the supreme sacrifice they made on our behalf. They experienced at first hand the horror of armed conflict. Yet since that time, we have now entered a new phase, a completely new phase, of problems about warfare. The advent of nuclear warfare has made us all re-examine in an agonising way the traditional theology of the just war . . .

The Catholic Church has repeatedly made known its longing for peace; it has expressed frequently its horror at the possibility of nuclear war and its desire for controlled but universal disarmament. Over a decade ago, the second Vatican Council declared 'Any act of war aimed indiscriminately at the destruction of entire cities or of extensive areas along with their population is a crime against God and man himself. It merits unequivocal and unhesitating condemnation' (*Gaudium et Spes*, n. 80). But the debate among Catholics proceeds vigorously. Does the Church mean that any and every use of nuclear weapons is ruled out and that a country may never threaten to use them against an aggressor?

Some uses of nuclear weapons would not involve the destruction of civilian populations; their use might be justified without too great moral difficulty. Examples are anti-ballistic missiles with nuclear warheads, air-to-air missiles, anti-submarine weapons. And it may be that escalation is not a certainty. It is possible to argue, as NATO does, that the discriminate non-strategic use of nuclear weapons has a finite chance of ending conflict by convincing an aggressor of one's resolve not to accept defeat and persuading the aggressors to desist to avoid unacceptable harm. It is then possible, at least in theory, to draw a distinction between having a nuclear deterrent to attack military objectives and to have one's deterrent aimed at the wholesale destruction of civilian targets. Let us consider the morality of these options.

I think it is quite clear from the authoritative teaching of the Second Vatican Council that the indiscriminate killing of civilian populations is immoral and can never be justified. This moral condemnation applies both to attack and counter-attack. Total war is immoral under all circumstances. It would also follow that it is wrong, in my view, to seek to deter an aggressor by threatening to wage total war in this morally unacceptable way ...

I could not, however, similarly condemn outright the possession of nuclear arms which are directed to military targets. But I would wish to emphasise two vital conditions. First, their possession and use would not be justified unless it is possible in practice to draw a clear distinction between military installations and personnel who will be destroyed and the civilian population which may be affected. And secondly, the use of strategic weapons of this type must not lead to escalation. If these two conditions do not obtain then it is very doubtful that even deterrent weapons directed to military targets can be morally justified.

Extract from *To Live in Christ Jesus*: National Conference of Roman Catholic Bishops of the United States, 1976[4]

The Church has traditionally recognised that, under stringent conditions, engaging in war can be a form of legitimate defence. But modern warfare, in both its technology and in its execution, is so savage that one must ask whether war as it is actually waged today can be morally justified.

At the very least all nations have a duty to work to curb the savagery of war and seek the peaceful settlement of disputes. The right of legitimate defence is not a moral justification for unleashing every form of destruction. For example, acts of war deliberately directed against innocent noncombatants are gravely wrong, and no one may participate in such an act. In weighing the morality of warfare today, one must also take into consideration not only its immediate impact but also its potential for harm to future generations: for instance, through pollution of the soil or the atmosphere or damage to the human gene pool...

With respect to nuclear weapons, at least those with massive destructive capability, the first imperative is to prevent their use. As possessors of a vast nuclear arsenal, we must also be aware that not only is it wrong to attack civilian populations but it is also wrong to threaten to attack them as part of a strategy of deterrence. We urge the continued development and implementation of policies which seek to bring these weapons more securely under control, progressively reduce their presence in the world, and ultimately remove them entirely.

Resolution passed at the British Council of Churches Assembly, November 1979

The Assembly of the British Council of Churches
1. (a) welcomes the publication of the Report 'The Future of the British Nuclear Deterrent';
   (b) thanks the Working Party of the Council on Christian Approaches to Defence and Disarmament;
   (c) urges Christians to study the issues raised in the Report.
2. (a) reaffirms the conviction which the Council expressed in 1963 that nuclear weapons 'are an offence to God and a denial of His purpose for man. Only the rapid progressive reduction of these weapons, their submission to strict international control and their eventual abolition can remove this offence. No policy which does not explicitly and urgently seek to realise these aims can be acceptable to Christian conscience';
   (b) believes that the non-replacement by the UK of its present nuclear strategic deterrent (the Polaris missile system) would strengthen moves for nuclear non-proliferation, and urges Her Majesty's Government to take a decision to this effect;
   (c) invites other Governments to take comparable confidence-building measures of restraint or renunciation, and encourages

Appendix 191

Christians actively to promote such steps.

Resolution passed at the British Council of Churches Assembly, November 1980

The Assembly registers the following:
1. The continuing escalation of nuclear arms threatens the very security which the weapons are held to guarantee.
2. The development and deployment of nuclear weapons has raised new and grave ethical questions for Christians. Because no gain from their use can possibly justify the annihilation they would bring about and because their effects on present and future generations would be totally indiscriminate as between military and civilians, to make use of the weapons would be directly contrary to the requirements of the so-called just war. The doctrine of deterrence based upon the prospect of mutual assured destruction is increasingly offensive to the Christian conscience.
3. The resources devoted to military expenditure of all kinds are desperately needed to tackle the world-wide and domestic problems of poverty, hunger, ignorance and disease. Military budgets make demands which are denials of Christian understandings of how resources are to be used.
4. The time has come for a more resolute involvement of Christians in the current debate about defence and disarmament and in the taking of new initiatives for peace. Such initiatives must include attempts to use and increase contacts between Christians in the West and in Eastern Europe and the Soviet Union.
5. Therefore this Assembly calls upon all Christians to support the World Disarmament Campaign as one way of advocating the multilateral or unilateral approach towards disarmament, by signing the petition and encouraging others to do so.

Extracts from the Archbishop of Canterbury's speech during the British Council of Churches debate on nuclear weapons, November 1980

It is impossible when talking of nuclear warfare to talk about the just war — the war prosecuted in a just cause in which gains can be judged to outweigh the inevitable injuries inflicted on individuals and societies. After the Second World War, even the defeated nations, Germany and

Japan, were able to recover fairly rapidly. This would not be true after a Third World War.

Nuclear warfare also makes it virtually impossible to draw distinctions between combatants and non-combatants. It may be possible to have a just war, but there can be no such thing as just mutual obliteration . . .

I have been struggling with my conscience about the question of unilateral disarmament. Could this be a new way to halt the spiral and begin to put it into reverse? There are many sincere people who believe that, if this country alone were to abandon its nuclear weapons, then this would be a potent example to others and also make it very unlikely that this country would be attacked in its turn by nuclear means.

I must say that I doubt the exemplary power of this gesture which would be made as we still expect for the foreseeable future to be sheltering under the American nuclear umbrella, and I fear that it might even serve to destabilize a balance which has contributed to the peace of Europe for thirty-five years. By unilaterally abandoning our nuclear deterrent and refusing to accommodate American missiles on our soil, we might further contribute to the further disintegration of the alliance between the United States and Europe which has been the basis of our security for the last four decades . . .

I feel that two of the most profitable ways of reducing tension and the dangers of the present situation would be some attempt to ban and scrap battle-field nuclear weapons which contribute to the illusion that limited and tactical nuclear warfare is possible, and also we ought to see whether it would not be possible to gain a pledge from each of the Nuclear Powers that they would not use nuclear weapons first.

We ought also to press forward with non-proliferation treaties and an extension of the test ban, and with further negotiations to limit chemical warfare, while realising that above all the world is dangerous because all regimes in varying degrees do not sufficiently respect the Word and deal in lies and propaganda which create the possibility of doing the unthinkable: destroying human life and our planet.

Notes

1. Extract from Encyclical Letter of Pope John XIII on Human Rights and Duties, *Pacem in Terris* (1963) (translation by Henry Waterhouse SJ, Catholic Truth Society, London, 1980).

2. *Gaudium et Spes* (1965), in W.M. Abbott SJ and J. Gallagher (eds.), *The Documents of Vatican II* (Geoffrey Chapman, London, 1966).

3. 'The Morality of Nuclear Defence': Cardinal Hume's address to the first national convention of the World Disarmament Campaign, held at Central Hall, Westminster, on 12 April 1980.

4. National Conference of Catholic Bishops of the United States, *To Live in Christ Jesus: A Pastoral Reflection on the Moral Life* (US Catholic Conference, Washington, 1976).

INDEX

Aden 126
Afghanistan, Soviet invasion of 20, 21, 23, 24, 26, 30, 59, 64, 70, 128, 139
Ambrose, St 45
Angola 25, 117
Antarctic Treaty 132
Anti-Ballistic Missile Treaty 33, 104, 107, 127, 135
anti-satellite activities 35, 123, 135-6
anti-submarine technology 72
Apel, Dr Hans 30
Apollo-Soyuz link-up 117
Aquinas, St Thomas 83
Arab-Israeli wars 25
Arab League 19
Argentina 35
arms control: need for 34-5; prospects for 139-41; technology and 197; *see also* multilateral disarmament, unilateral disarmament
arms industry 64
arms race: Christianity and 53; impetus for 33, 136
Association of South East Asian Nations (ASEAN) 19
Augustine, St 45
Ayacucho, Declaration of 126

Backfire bomber 21, 32, 123, 128
Bailey, Sidney D. 62
Baluchistan 23
Baruch Plan 122
battlefield nuclear weapons 34, 73, 192
Beaufre, General André 71
Berbera 126
Berlin, West 16, 23, 25, 30, 96, 105, 114-15
Bible, the 41-5 *passim*, 47, 51, 55, 167, 169
biological weapons 118, 132-3
Biological Weapons Convention 117, 133
Bishops, National Conference of (USA) 189-90

Bismarck, Prince Otto von 153
booby traps 134, 162
Brazil 35, 134
Brezhnev, President Leonid 59, 118, 127, 134
British Council of Churches 13, 62-3, 120, 181, 190-1
Brown, Harold 82
Brzesinski, Z. 28
Bundy, McGeorge 32, 81
Butterfield, Herbert 36

Campaign for Nuclear Disarmament 121
Canada 62
cancer 81
Carter, President Jimmy (James Earl, Jnr) 59, 113, 123, 127, 128
chemical weapons 89, 122-3, 132, 133
Chevaline programme 60
Chile 59
China 26, 60, 146; nuclear weapons testing 130, 131
Churches, statements on war and disarmament 185-92
civil defence 66
Committee on Disarmament (CD) 125, 127, 132, 133, 139, 140
Commonwealth of Socialist States 27
Conference of Communist Parties 27
Conference on Security and Co-operation in Europe (CSCE) 29, 117; Final Act 122; Review Conference 134
confidence building measures 29, 34, 64, 117, 134
conflicts *see* wars and conflicts
conventional weapons: casualties of 88-9, *see also* under wars and conflicts; control of 133-4; *see also* war, laws of
Council on Christian Approaches to Defence and Disarmament (CCADD) 13, 14, 51, 166, 190
cruise missiles in Europe 32, 60, 64, 66, 86, 129

Index

Cuba 24; missile crisis 16, 91, 105, 117
Cyprian, St 56
Czechoslovakia 20, 59, 137

Denmark 134
détente: arms control and disarmament and 31-5; Christians and 31; criticisms of 29; necessity for 28-35; Soviet perceptions of 29-30
deterrence: balance of 69-70, 75-9, 176-9; Christianity and 36, 41, 85, 180; conventional weapons and 74-5; cost of 112-13; credibility and 105-6; defence of 68-91; definition of 16, 25, 28-9; delicacy 97, 106, 176-8; détente, relationship with 16; disarmament and 119-21; ethical considerations 13, 79-88 *passim*, 171-5; intention and 50, 80-1, 83-4, 85, 97-9, 171-2, 173-5; nuclear weapons 71, 72, 74; objectives 80; requirements for 70-1; robustness of 16, 103, 104, 106, 108-11 *passim*, 176-8; size of 71-2; use, will to 50; world poverty and 112-13
disarmament: obstacles to 136-8; *see also* multilateral disarmament, unilateral disarmament
Disarmament Commission 134
Disarmament Committee 130
Disarmament Decade 126
double effect, principle of 83
Dresden 81

East-West relations 17, 19, 57-61, 169-71, 181, 182; power and ideology and 21-5
Egypt 24
Eighteen-Nation Disarmament Committee 130-1
Ethiopia 23, 24, 25, 117
Europe: balance of forces in 16-17, 20-3; conventional forces in 21, 60, 121; nuclear free zone in 65, 121; nuclear weapons in 21-2, 34, 60, 73, 123, 128-9, 137, cruise missiles 32, 60, 64, 66, 86, 129; Soviet Union, fears of invasion by 22, 170, 172-3, 175; war in 95, 97, 129, *see also* First World War,

Second World War; *see also* Conference on Security and Cooperation in Europe, Mutual and Balanced Force Reduction Talks
European Nuclear Disarmament campaign 65, 121

F-111 128
faith 46, 47
Ferguson, Professor John 56-7
first strike 34, 72, 177, 178
First World War 68, 69
'flexible response' 73
France 69, 131, 136; disarmament proposals 34, 134; nuclear weapons testing 130
Franco-Prussian War 153-4
'fratricide' 72

Gambetta, Leon 153
Gaudium et Spes 186-7, 188
Geneva Conventions 145-8 *passim*, 155, 159, 160-1, 163, 183; additional protocols to 134, 161-2, 163
Germany, division of 29
Germany, East 158
Germany, West 29, 30
Giap, General 154
Goodwin, Geoffrey 13, 166
Greece 22
Gulf states 23-4, 25

Hague Law 147, 150-1, 161
Hamburg 81
Harmel Report 30
Helsinki Conference *see* Conference on Security and Co-operation in Europe
Hill-Norton, Lord 84
Hiroshima 68, 81
Hitler, Adolf 68-9, 102
Hope-Jones, Ronald 13, 14
Howard, Michael 28, 32, 66
Huber, Max 148, 150, 168
humanitarianism 148-50; cultural values and 156-60 *passim*
human rights 147
Hume, Cardinal 81, 173, 188-9
Hungary 20, 137

incendiary weapons 134, 162
India 20, 35, 130, 131, 134, 158
Indian Ocean 126

196 *Index*

Indonesia 159
inhumane weapons, UN conference on 134, 161, 162, 182, 183
inter-continental ballistic missiles (ICBMs) 21-2, 72, 128
Interim Agreement on the Limitation of Strategic Offensive Arms 117
Intermediate Range Ballistic Missiles 21; *see also* Pershing II, SS.20
International Atomic Energy Agency 35, 131, 132
International Committee of the Red Cross 146-7, 155, 160, 161-3 *passim*, 183
International Institute for Strategic Studies (IISS) 22, 72
International Satellite Monitoring Agency 136
international society, notion of 17-18
Iran 24, 26, 156
Iraq 35, 131
Ireland, Northern 64
Ireland, Republic of 130
'Islamic bomb' 131
Islamic fundamentalism 18, 26
Israel 21, 35, 131

Japan 26, 71; in Second World War 68, 69, 79, 81, 156
Jesus Christ 31, 42-3, 53
John XXIII, Pope 56, 185
John Paul II, Pope 63, 187-8
Johnson, President Lyndon 29

Kampuchea 20, 59
Kennan, George 58, 75
Kennedy, President John F. 105
King-Hall, Sir Stephen 64
Kissinger, Henry 109
Korea, South 59
Kortunov, Professor V. 29

Lambeth Conferences 40, 185
laser beams 135
Lateran Council, Second 68
Law of the Sea, UN Commission on 179
laws of war *see* war, laws of
Legault, Albert 75-9, 178
liberation movements 20, 146, 163
liberation theology 41
Libya 156
Lindsay of Birker, Lord 84
Lindsey, George 75-9, 178

Luke, St 43

MacDonald, Ramsay 122
MacKinnon, Donald 28
McNamara, Robert 72, 82
Malaysia 159
Manichaeism 28, 57-61
Martens, F.F. 151
Medvedev, Roy 122
military expenditure 107, 123, 126, 191
Miller, Richard 158, 160
mines 134, 162
MIRVs (multiple independently-targeted re-entry vehicles) 77
Moore, G.E. 84
Moscow 60
Mountbatten, Lord Louis 64-5, 124
Mozambique 25
multilateral disarmament: arguments for 100-1; background to 117-18, 122-3; definition of 100; ethical aspects 118-19; negotiating 123-6, 179-80; past experience of 122-3; unilateral disarmament and 13, 61, 100-12, 124, 176-9, 180, 181; *see also under names of treaties and agreements*
Munro, H.H. 157
Mutual and Balanced Force Reduction talks (MBFR) 23, 117, 134

Nagasaki 81
Namibia 21
napalm 134, 162
nationalism 55-6
NATO (North Atlantic Treaty Organization): armed forces 134; conventional weapons 21, 33, 114; military expenditure 123; nuclear weapons: levels of 95-6, no-first-use pledge 95, 114, 172, reasons for deploying 95-6, 114, strategic 60, 95, theatre 32, 33-4, 95, 123, 128-9; strategic doctrines 33, 95, *see also under* United States of America; UK's withdrawal from 65; unity of 22; *see also under names of member states*
Nehru, Pandit 129
neutrality 120
Nigeria 159

Index

Non-Proliferation Treaty 62, 107, 117, 127, 130-2 *passim*, 139; Review Conference 131-2
nuclear power, peaceful use of 35, 132
Nuclear Suppliers Group 132
nuclear war: conventional war and 66; deterrence and 98, 99; escalation of 74, 95, 105; fighting conditions 73; fighting methods 33; 'limited' 33, 64-5; non-combatants and 94; risk of 16
nuclear weapons: blackmail and 88, 107-8; detonation of 72; development of 69; fallout 94; 'fratricide' 72; indiscriminate killing by 81-4, 94, 189; proliferation of 35, 130-2; radiation 81; reduction of numbers of 94-115 *passim*; tests, stopping 129-30, *see also* Partial Test Ban Treaty; use of 74; yields of 73-4
'Nuremberg principles' 160

oil 23, 168
Ollivier, Émile 69
Organization of African Unity (OAU) 19
Organization of American States (OAS) 19
Osgood, Charles 62
Outer Space Treaty 33, 132, 135

Pacem in Terris 46, 56, 185-6
Pakistan 20, 23, 35, 130, 131, 158
Palestine Liberation Organization (PLO) 21, 24
Parrent, Allan 84
Partial Test Ban Treaty 62, 104, 107, 117, 129-30, 139
Paul, St 42, 43, 55
Pax Christi 63
peace movement 121-2
Pershing II 32, 129
Pictet, Jean 148, 150, 151, 155, 156, 168
Plumer, Field Marshal 74
Poland 20, 26, 137
Polaris 60, 86, 177, 190
politics, Christianity and 41-51 *passim*
Pol Pot 20, 59
prophecy 47
public opinion 59-60, 62, 63, 111

'Racek, Vaclav' 122
radiological weapons 132, 133
rapid deployment force 23-4
Reagan, President Ronald 118, 128, 139
Red Cross 145, 154-6, 164; *see also* International Committee of the Red Cross
research and development 135
Röling, B.V.A. 163
Ross, Sir David 84-5, 175
Runcie, Archbishop 120, 191-2
Russia *see* Union of Soviet Socialist Republics

St Petersburg Declaration 162
'Saki' 157
SALT (Strategic Arms Limitation Talks) 35, 59, 62, 117, 127-8, 135, 140; *see also* Anti-Ballistic Missile Treaty, Interim Agreement on the Limitation of Strategic Offensive Arms, Vladivostok Accord
satellites 136; *see also* anti-satellite activities
Schlesinger, James 113
Seabed Treaty 132
second strike capability 25
Second World War 68, 69, 88, 89, 145, 146, 156, 160
Shelley, Percy Bysshe 87
SIPRI (Stockholm International Peace Research Institute) 123, 129, 170
Somalia 24
Sonnenfeldt, Helmut 29
South Africa 131
space, arms race in 123; *see also* Outer Space Treaty
Space Shuttle 136
SS.20 missiles 21, 32, 66, 121, 123, 128, 129
Stalin, Josef 147, 157
submarine-launched ballistic missiles (SLBMs) 22; *see also* Polaris, Trident
submarines, ballistic-missile 72
Suez 20, 30
Sweden 133,
Synod of Bishops, Roman 54

Tansley Report 156
Tanzania 20

Tàssigny, General de Lattre de 157
Taylor, A.J.P. 153
technological advance 34, 35, 75, 130, 135-6, 160, 161, 177
theology 46-51 *passim*
Thompson, E.P. 65, 102, 122
Tlatelolco, Treaty of 132
Trident 34, 86, 87, 179, 190
Turkey 22

Uganda 20
unilateral disarmament: arguments against 103-15 *passim*, 122, 123-4, 140, 180, 192; arguments for 101-3, 181; arms control and 106-7; Christianity and 53-66 *passim*; conventional weapons and 114; definition of 100; differences among supporters of 179; global strategies 108-10; measures for 63-4; multilateral disarmament and 13, 61, 100-12, 124, 176-9, 180, 181; NATO and 109; public opinion 62; response to 111-12; risks of 111-12; strategy and 100, 102-3, 114-15
Union of Soviet Socialist Republics: anti-satellite activities 135; armed forces, numbers of 123; chemical weapons 21, 123; China, relations with 60; conscription 123; cruise missiles, attitude to 66; disarmament and 127, 134, 137, 138; Eastern Europe, control of by 26-7; Europe, conventional forces in 21; Europe, invasion of by 22, 170, 172-3, 175; foreign adventures of 20, 21, 24, 91, 103, 117, 126; ideology of 27-8, 138-9; law of war and 146; military expenditure 137, 139; military power, build-up of 21, 32, 170; Muslim republics 26; nuclear weapons 22, 137; perceptions of the world 26-7; political image of 25; protest stifled by 121-2; Second World War, casualties in 69; strategic doctrines of 90, 94-5, 97, 105; unilateral disarmament and 121; United States of America, relations with 57-60; world role of 19, 24-5
United Kingdom: chemical weapons 122-3, 133; conscription ended 123; conventional forces of 34; cruise missiles in 60, 64, 66, 86; Defence White Paper (nuclear weapons), 1981 60, 71, 88-91, 119-21; disarmament proposals 133, 138; Geneva Conventions and 161-2; government propaganda 59-60; information, manipulation of 60; military expenditure 123, 137; NATO and 65, 86, 90, 120, 179; nuclear forces of 34, 60, 64, 87, 174, 177, 190, *see also* Polaris, Trident; nuclear missiles targeted on 120; Second World War, casualties in 69; unilateral disarmament of 63-4, 120, 121
United Nations: charter of 20, 124-5, 133; disarmament and 126, 127, 131-2, 133, 136, 137-8; General Assembly 19, 125, 126, 127, 130; Inhumane Weapons, Conference on 134, 161, 162, 182, 183; international relations and 19; Law of the Sea Commission 179; peacekeeping forces 20, 21, 138; Security Council 19, 125; *see also following entries*
United Nations Emergency Force II (UNEF II) 21
United Nations Interim Force in Lebanon (UNIFIL) 21
United Nations Special Session on Disarmament (UNSSD) 125, 126, 133, 136, 139, 140; Final Document of 61, 63, 125, 126; Report 66
United States of America: anti-satellite activities 135-6; armed forces, strength of 22, 123; arms transfers 123; conscription 123; disarmament initiatives 59, 122, 123, 127, 129; military expenditure 123, 135; nuclear weapons 21-2, 32, 95, 96, 123, 128-9; Presidential Directive 59 33, 82, 113; research and development 135; strategic doctrines of 82, 94-7 *passim*, 103, 105, 109, 113-14; Union of Soviet Socialist Republics, relations with 57-60
Urban II, Pope 37

Index

Vatican Council, Second 81, 82, 188, 189
Venables, Mark 14
Vietnam 20, 24
Vietnam, North 146, 154
Vietnam War 20, 30, 59, 64, 69, 123
violence, Christianity and 56
Vladivostok Accord 117, 127
Vulcan aircraft 128

Walzer, Michael 87-8
warfare: attitudes to 152-4; changes in 68-9, 143-4, 182; Christians and 40-51 *passim*, 150-1, 166-9; historical perspective on 68-9, 152-4, 182; Marxist-Leninist and 154; nuclear weapons and 69; *see also following entries and* nuclear war
wars: casualties of 69, 88-9, 143-4, 182; just, concept of 41, 44, 49-50, 80-1, 87; numbers of 127, 182; superpowers' intervention in local 18-19, 20; world significance of 18-21 *passim; see also following entry*
war, laws of: civilians and 156, 157; cultural values 156-60 *passim*; difficulties of 162-3; indifference to 182-3; origins of 144; purpose of 145; soldiers and 156, 157; theory of 145-6; trust and 152
Warsaw Treaty Organization: armed forces of 134; conventional weapons 172; military expenditure 123; *see also under names of member states*
Weber, Theodore 88
World Council of Churches 41
World Disarmament Campaign 173, 188
World Disarmament Conference 122, 137, 191

Yemen, South 23

Zaire 159
Zimbabwe 20
Zuckerman, Lord 135

For Product Safety Concerns and Information please contact our EU representative GPSR@taylorandfrancis.com
Taylor & Francis Verlag GmbH, Kaufingerstraße 24, 80331 München, Germany

www.ingramcontent.com/pod-product-compliance
Lightning Source LLC
Chambersburg PA
CBHW070302010526
44108CB00039B/1587